LibreCAD 2.2 Black Book

By
Gaurav Verma

Rohan Sharma

(CADCAMCAE Works)

& Mr. Anutosh

Edited by
Kristen

ISBN # 978-1-77459-109-3

NOTICE TO THE READER

Publisher does not warrant or guarantee any of the products described in the text or perform any independent analysis in connection with any of the product information contained in the text. Publisher does not assume, and expressly disclaims, any obligation to obtain and include information other than that provided to it by the manufacturer.

The reader is expressly warned to consider and adopt all safety precautions that might be indicated by the activities herein and to avoid all potential hazards. By following the instructions contained herein, the reader willingly assumes all risks in connection with such instructions.

The Publisher makes no representation or warranties of any kind, including but not limited to, the warranties of fitness for a particular purpose or merchantability, nor are any such representations implied with respect to the material set forth herein, and the publisher takes no responsibility with respect to such material. The publisher shall not be liable for any special, consequential, or exemplary damages resulting, in whole or part, from the reader's use of, or reliance upon, this material.

DEDICATION

To teachers, who make it possible to disseminate knowledge
to enlighten the young and curious minds
of our future generations

To students, who are the future of the world

THANKS

To Mr. Anutosh for the contribution in the book.

To my friends and colleagues

To my family for their love and support

Training and Consultant Services

At CADCAMCAE WORKS, we provide effective and affordable one to one online training on various software packages in Computer Aided Design(CAD), Computer Aided Manufacturing(CAM), Computer Aided Engineering (CAE), Computer programming languages(C/C++, Java, .NET, Android, Javascript, HTML and so on). The training is delivered through remote access to your system and voice chat via Internet at any time, any place, and at any pace to individuals, groups, students of colleges/universities, and CAD/CAM/CAE training centers. The main features of this program are:

Training as per your need

Highly experienced Engineers and Technician conduct the classes on the software applications used in the industries. The methodology adopted to teach the software is totally practical based, so that the learner can adapt to the design and development industries in almost no time. The efforts are to make the training process cost effective and time saving while you have the comfort of your time and place, thereby relieving you from the hassles of traveling to training centers or rearranging your time table.

Software Packages on which we provide
basic and advanced training are:

CAD/CAM/CAE: CATIA, Creo Parametric, Creo Direct, SolidWorks, Autodesk Inventor, Solid Edge, UG NX, AutoCAD, AutoCAD LT, EdgeCAM, MasterCAM, SolidCAM, DelCAM, BOBCAM, UG NX Manufacturing, UG Mold Wizard, UG Progressive Die, UG Die Design, SolidWorks Mold, Creo Manufacturing, Creo Expert Machinist, NX Nastran, Hypermesh, SolidWorks Simulation, Autodesk Simulation Mechanical, Creo Simulate, Gambit, ANSYS and many others.

Computer Programming Languages: C++, VB.NET, HTML, Android, Javascript and so on.

Game Designing: Unity.

Civil Engineering: AutoCAD MEP, Revit Structure, Revit Architecture, AutoCAD Map 3D and so on.

We also provide consultant services for Design and development on the above mentioned software packages

For more information you can mail us at:
cadcamcaeworks@gmail.com

Table of Contents

Chapter 2 : Sketching

Chapter 3 : Advanced Dimensioning and Practice

Chapter 4 : Practical and Practices

Preface

LibreCAD is an open-source parametric 2D Computer Aided Design (CAD) application, made primarily to design 2 dimensional objects and drawings. Current version of LibreCAD is based on Qt5 GUI libraries which supports multiple platforms. So, LibreCAD can run on operating systems like Microsoft Windows, Macintosh, and Unix. Also, LibreCAD is available in more than 30 languages like English, German, Hindi, Arabic, French, Russian, and so on. You can download your free copy of software from *https://wiki.librecad.org/index.php/Download*

The **LibreCAD 2.2 Black Book** is the 1st edition of our series on LibreCAD. This book is written to help beginners in creating various 2D geometries and drawings related to different fields. The book follows a step by step methodology. In this book, we have tried to give real-world examples with real challenges in drafting. The book covers almost all the information required by a learner to master the LibreCAD. Some of the salient features of this book are:

In-Depth explanation of concepts

Every new topic of this book starts with the explanation of the basic concepts. In this way, the user becomes capable of relating the things with real world.

Topics Covered

Every chapter starts with a list of topics being covered in that chapter. In this way, the user can easily find the topics of his/her interest easily.

Instruction through illustration

The instructions to perform any action are provided by maximum number of illustrations so that the user can perform the actions discussed in the book easily and effectively. There are about 480 illustrations that make the learning process effective.

Tutorial point of view

At the end of concept's explanation, the tutorial make the understanding of users firm and long lasting. Almost each chapter of the book has tutorials that are real world projects. Moreover most of the tools in this book are discussed in the form of tutorials.

For Faculty

If you are a faculty member, then you can ask for video tutorials on any of the topic, exercise, tutorial, or concept. As faculty, you can register on our website to get electronic desk copies of our latest books. Faculty resources are available in the **Faculty Member** page of our website (**www.cadcamcaeworks.com**) once you login. Note that faculty registration approval is manual and it may take two days for approval before you can access the faculty website.

Formatting Conventions Used in the Text

All the key terms like name of button, tool, drop-down etc. are kept bold.

Free Resources

Link to the resources used in this book are provided to the users via email. To get the resources, mail us at **cadcamcaeworks@gmail.com** with your contact information. With your contact record with us, you will be provided latest updates and informations regarding various technologies. The format to write us mail for resources is as follows:

Subject of E-mail as **Application for resources of .... book**.
Also, given your information like
Name:
Course pursuing/Profession:
Contact Address:
E-mail ID:

Note: We respect your privacy and value it. If you do not want to give your personal informations then you can ask for resources without giving your information.

About Authors

The author of this book, Gaurav Verma, has written and assisted in more than 17 titles in CAD/CAM/CAE which are already available in market. He has authored Autodesk Fusion 360 Black Book, AutoCAD Electrical Black Book, Autodesk Revit Black Books, and so on. He has provided consultant services to many industries in US, Greece, Canada, and UK. He has assisted in preparing many Government aided skill development programs. He has been speaker for Autodesk University, Russia 2014. He has assisted in preparing AutoCAD Electrical course for Autodesk Design Academy. He has worked on Sheetmetal, Forging, Machining, and Casting designs in Design and Development departments of various manufacturing firms. If you have any query/doubt in any CAD/CAM/CAE package, then you can contact the author by writing at cadcamcaeworks@gmail.com

For Any query or suggestion

If you have any query or suggestion, please let us know by mailing us on *cadcamcaeworks@gmail.com*. Your valuable constructive suggestions will be incorporated in our books and your name will be addressed in special thanks area of our books on your confirmation.

Chapter 1

Starting with LibreCAD

The major topics covered in this chapter are:

- *Overview of LibreCAD*
- *Installing LibreCAD*
- *Starting LibreCAD*
- *File Menu*
- *View Menu*
- *Standard Views*
- *Freeze Display*
- *Draw Style*
- *Tools Menu*
- *Navigating in the 3D view*
- *FreeCAD Interface*

OVERVIEW OF LIBRECAD

LibreCAD is an open-source parametric 2D Computer Aided Design (CAD) application, made primarily to design 2 dimensional objects and drawings. Parametric modeling is a certain type of modeling, where the shape of the 2D objects you design is controlled by parameters. For example, the shape of a rectangle can be controlled by two parameters: width and length. Current version of LibreCAD is based on Qt5 GUI libraries which supports multiple platforms. So, LibreCAD can run on operating systems like Microsoft Windows, Macintosh, and Unix. Also, LibreCAD is available in more than 30 languages like English, German, Hindi, Arabic, French, Russian, and so on. Visit the link "https://librecad.readthedocs.io/en/latest/appx/languages. html" to get more information about supported languages. The first view of interface is shown in Figure-1. Various file formats supported by LibreCAD are given next.

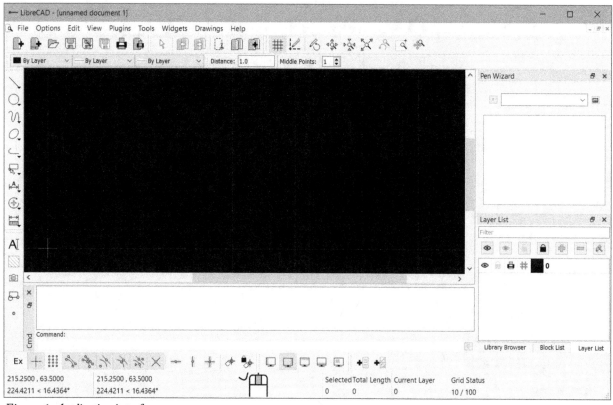

Figure-1. Application interface

Open File or Import Block
CAD : DXF, DWG, JWW
CAD font : LFF, CXF

Import Image
Vector image : SVG, SVGZ
Bitmap image : BMP, CUR, GIF, ICNS, ICO, JPEG, JPG, PBM, PGM, PNG, PPM, TGA, TIF, TIFF, WBMP, WEBP, XBM, XPM

Save File
CAD : DXF (2007), DXF (2004), DXF (2000), DXF (R14), DXF (R12)
CAD font : LFF, CXF

Export
PDF

Vector image : SVG, CAM (Plain SVG)
Bitmap image : BMP, CUR, ICNS, ICO, JPG, PBM, PGM, PNG, PPM, TIF, WBMP, WEBP, XBM, XPM

LibreCAD is not designed for a particular kind of work or to make a certain kind of objects. Instead, it allows a wide range of uses and permits users to produce drawings of all sizes and purposes from small electronic components to all the way up to buildings. Each of these tasks have different dedicated sets of tools and workflow available. Being open-source, LibreCAD benefits from the contributions and efforts of a large community of programmers, enthusiasts, and users worldwide.

LibreCAD also benefits from the huge, accumulated experience of the open-source world. It also possesses all kinds of features that have become a standard in the open-source world, such as supporting a wide range of file formats being hugely scriptable, customizing, and modifiable. All this is made possible through a dynamic and enthusiast community of users.

DOWNLOADING AND INSTALLING LIBRECAD
- Connect your PC with the internet connection and then go to the website **https://librecad.org** as shown in Figure-2.

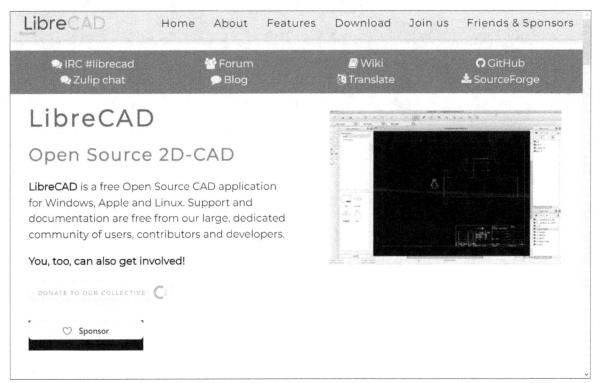

Figure-2. FreeCAD website

- Scroll down on the web page and select **from SourceForge** or **from GitHub** link button for desired platform; refer to Figure-3.

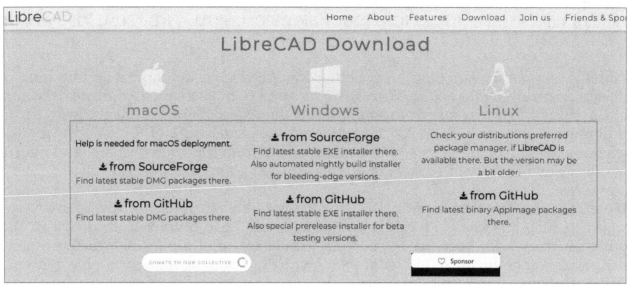

Figure-3. Download tab of FreeCAD website

- We are using `from GitHub` option for `Windows` platform. The LibreCAD releases page will be displayed; refer to Figure-4.

Figure-4. LibreCAD github releases

- Select desired release link button to get the setup files. On selecting the link button, the web page for downloading installer files will be displayed; refer to Figure-5.

Figure-5. LibreCAD installer file

- Select the LibreCAD Installer exe file to download it. The software will begin to download.
- Once downloading is complete, open the download setup file and follow the instructions as per the setup window. The software will be installed in a couple of minutes.

STARTING LIBRECAD

- To start **LibreCAD** from **Start** menu, click on **LibreCAD** folder in the **Start** menu and select the **LibreCAD** icon; refer to Figure-6.

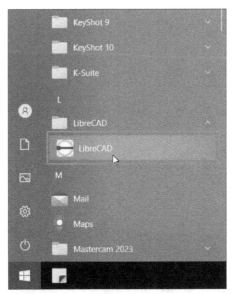

Figure-6. Start menu

- On starting the application, the initial screen of the application will be displayed as shown in Figure-7.

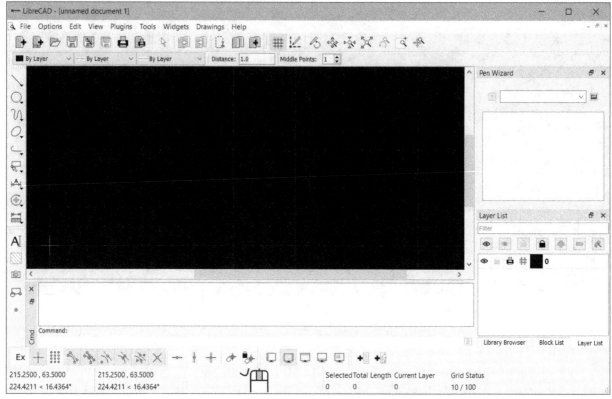

Figure-7. Initial screen of FreeCAD

FILE MENU

The options in the **File** menu are used to manage files and related parameters; refer to Figure-8. Various tools of **File** menu are discussed next.

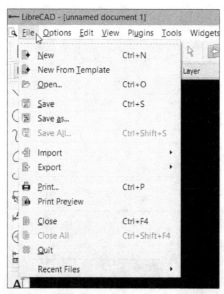

Figure-8. File menu

New

The **New** tool is used for initiating a new document file. The procedure to use this tool is discussed next.

- Click on the **New** tool from the **File** menu or **Toolbar**, or press **Ctrl+N** key from keyboard; refer to Figure-9. The new document file will open; refer to Figure-10. Note that we have changed the background color of drawing area to white for printing purpose. You will learn about the procedure to change color of drawing area later in this chapter.

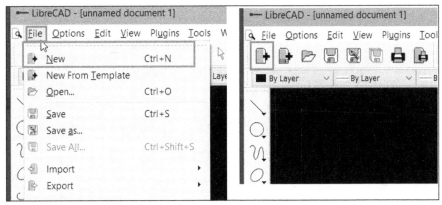

Figure-9. New tool

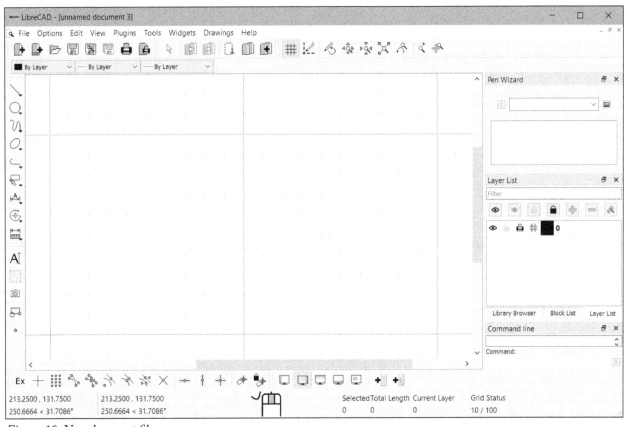

Figure-10. New document file

New From Template

The **New From Template** tool is used to create a new drawing file from a template. The procedure to use this tool is discussed next.

- Click on the **New From Template** tool from the **File** menu or **Toolbar**; refer to Figure-11. The **Open Drawing** dialog box will be displayed; refer to Figure-12.

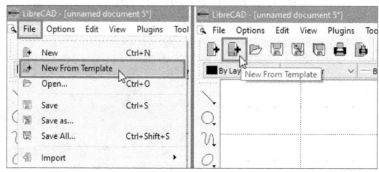

Figure-11. New From Template tool

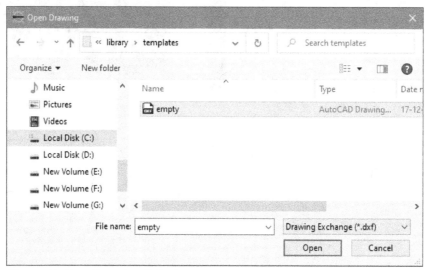

Figure-12. Open Drawing dialog box

- Select the default template, **"empty.dxf"**, that is installed with the application and found in the resources directory. Note that a user defined template can also be used instead of the default template by specifying the new template in the **Application Preferences**.

- After selecting the template, click on the **Open** button from the dialog box. A new document file will be opened as shown in Figure-10.

Open

The **Open** tool is used to open LibreCAD files and import files of other CAD applications. The procedure to use this tool is discussed next.

- Click on the **Open** tool from the **File** menu or **Toolbar** or press **CTRL+O** key from keyboard; refer to Figure-13. The **Open drawing** dialog box will be displayed; refer to Figure-14.

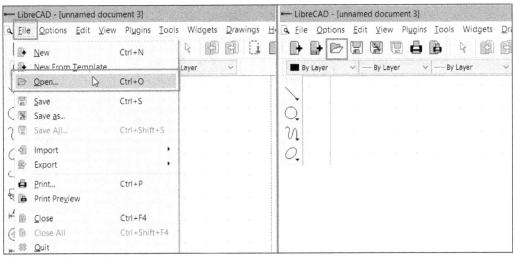

Figure-13. Open tool

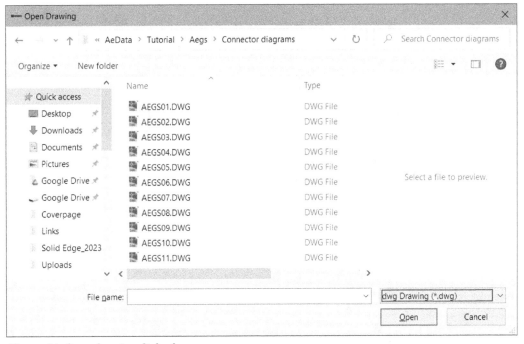

Figure-14. Open drawing dialog box

- Select desired file format from the **File Type** drop-down to define format of file to be opened. LibreCAD supports DWG, DXF, LFF, CXF, and JWW formats for opening files.
- Select desired file which you want to open and click on **Open** button from the dialog box. The file will open.

Close

The **Close** tool is used to close the current file without closing the application. The procedure to use this tool is discussed next.

- Click on the **Close** tool from **File** menu or press **Ctrl+F4** key from keyboard; refer to Figure-15. The current file will be closed.

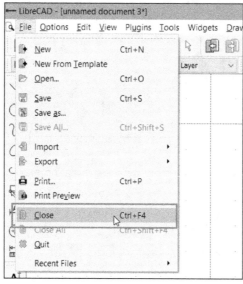

Figure-15. Close tool

Close All

The **Close All** tool is used to close all the opened files without closing the application. The procedure to use this tool is discussed next.

• Click on the **Close All** tool from the **File** menu or press **CTRL+SHIFT+F4** keys; refer to Figure-16. All the files which are currently open will be closed.

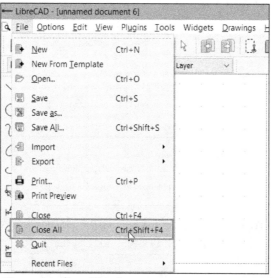

Figure-16. Close All tool

Save

The **Save** tool is used to save the current file opened in the application. The procedure to use this tool is discussed next.

• Click on the **Save** tool from **File** menu or **Toolbar** or press **Ctrl+S** key from keyboard; refer to Figure-17. The **Save Drawing As** dialog box will be displayed; refer to Figure-18.

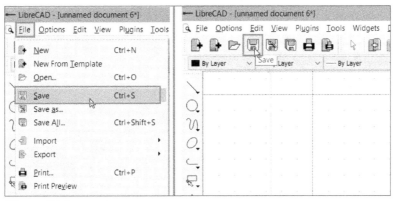

Figure-17. Save tool

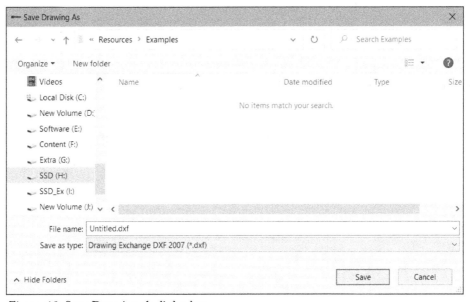

Figure-18. Save Drawing As dialog box

- Select desired format from the **Save as type** drop-down to define the type of file you want to save.
- Specify desired location and name for the file, and click on the **Save** button from the dialog box to save the file.

Save As

Using the **Save as** tool, you can save the file with different name. The procedure to use this tool is discussed next.

- Click on **Save as** tool from **File** menu; refer to Figure-19. The **Save Drawing As** dialog box will be displayed as shown in Figure-18.

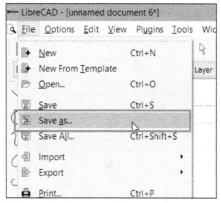

Figure-19. Save as tool

- Rest of the procedure is same as discussed in the **Save** tool.

Save All

The **Save All** tool is used to save all the documents opened in the application. The procedure to use this tool is discussed next.

- Click on the **Save All** tool from the **File** menu; refer to Figure-20. The **Save Drawing As** dialog box will be displayed as shown in Figure-18.

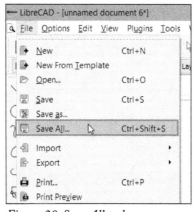

Figure-20. Save All tool

- Rest of the procedure is same as we have discussed earlier.

Importing Image/Block

Many times, importing reduces lots of extra work of rebuilding the base sketch for models. In LibreCAD, we can directly use the CAD files for creating or manipulating the model. The procedure to use this tool is discussed next.

- Select the **Insert Image** tool from the **Import** cascading menu; refer to Figure-21 to import an image in the drawing area. Similarly, select the **Block** option from the cascading menu to import drawing block. On selecting desired tool, respective dialog box will be displayed; refer to Figure-22 (in case of Image import).

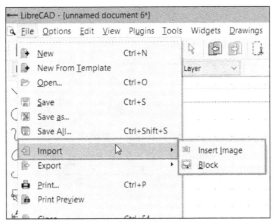

Figure-21. Import tool

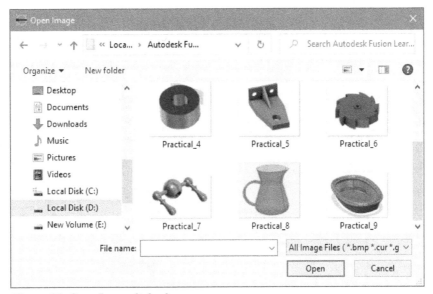

Figure-22. Open Image dialog box

- Select desired image to be inserted from the dialog box and click on the **Open** button. The **Tool Options** will be displayed; refer to Figure-23 and the selected image attached to cursor. You will be asked to specify the reference point.

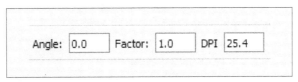

Figure-23. Insert Image Tool Options

- Specify desired angle value, scale factor, and DPI of the image to insert the image in the **Angle**, **Factor**, and **DPI** edit boxes of the **Tool Options**, respectively.
- After specifying desired parameters, click in the Drawing Window at desired location to insert the image. The image will be inserted; refer to Figure-24.

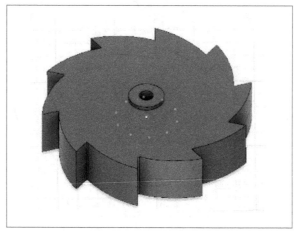

Figure-24. Image inserted

Exporting Drawing

The tools in the **Export** cascading menu are used to export current drawing file into various different formats. Various tools of this menu are discussed next.

Exporting as CAM or Plain SVG file

*   Click on the **Export as CAM/plain SVG** tool from **Export** cascading menu of the **File** menu; refer to Figure-25. The **Export as CAM/plain SVG** dialog box will be displayed; refer to Figure-26.

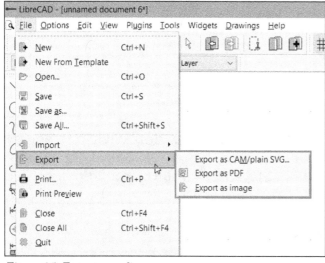

Figure-25. Export cascading menu

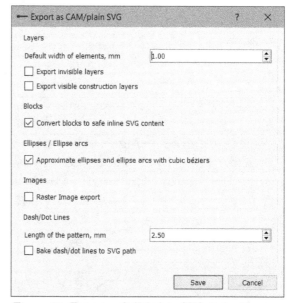

Figure-26. Export as CAM plain SVG dialog box

*   Specify desired width of lines in the **Default width of elements, mm** edit box.
*   Select the **Export invisible layers** check box to export objects on invisible layers as well.
*   Select the **Export visible construction layers** check box to export visible construction objects.
*   Select the **Convert blocks to safe inline SVG content** check box to convert drawing blocks into images when exporting.
*   Select the **Approximate ellipses and ellipse arcs with cubic beziers** check box to convert ellipses and elliptical arcs to splines when exporting.

- Select the **Raster Image export** check box to export images and vector graphics in drawing area as raster images.
- Specify desired value in the **Length of the pattern, mm** edit box to define the scale at which dash/dot patterns will be created in a dashed/dotted line.
- Select the **Bake dash/dot lines to SVG path** check box to convert dash/dot lines to full lines so that they can be later used for CAM toolpaths.
- After setting desired parameters, click on the **Save** button. The **Export as** dialog box will be displayed; refer to Figure-27.

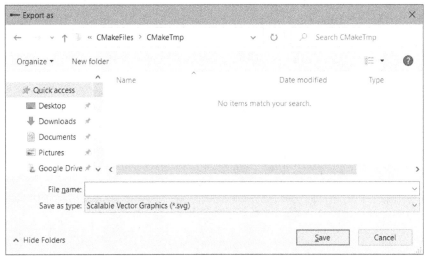

Figure-27. Export as dialog box

- Specify desired name of file in the **File name** edit box and click on the **Save** button.

Exporting as PDF

The **Export as PDF** tool is used to export current drawing to a PDF. The procedure to use this tool is given next.

- Click on the **Export as PDF** tool from the **Export** cascading menu of the **File** menu. The **Export as PDF** dialog box will be displayed; refer to Figure-28.

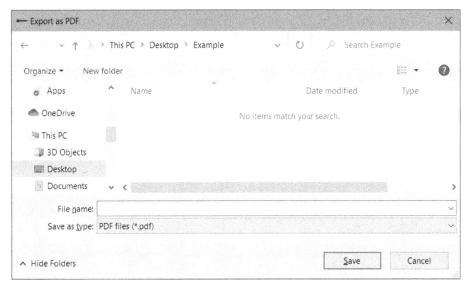

Figure-28. Export as PDF dialog box

- Specify desired name of file in the **File name** edit box and click on the **Save** button.

Exporting as Image

The **Export as image** tool is used to export current drawing as an image file. The procedure to use this tool is given next.

- Click on the **Export as image** tool from the **Export** cascading menu of the **Export** menu. The **Export as** dialog box will be displayed as discussed earlier.
- Specify desired name of file in the **File name** edit box.
- Select desired format for file from the **Save as type** drop-down in the dialog box and click on the **Save** button to save the image file. The **Image Export Options** dialog box will be displayed; refer to Figure-29.

Figure-29. Image Export Options dialog box

- Specify desired value of resolution (pixels per inch - ppi) in the **Resolution** edit box. The width and height of image will be decided automatically.
- Specify desired width and height of image in the **Width** and **Height** edit boxes, respectively. The resolution of image will be decided automatically.
- Specify desired value in **Left/Right - Border** edit box to define thickness of border line for image.
- Select the **set same size** check box to keep the border thickness same for top and bottom of image. Clear this check box to set different border thickness.
- Select desired radio buttons from the **Background** and **Colouring** areas to define colors of background and drawing objects, respectively.
- Click on the **OK** button from the dialog box to save the file.

Printing Drawing

The **Print** tool in **File** menu is used to print current drawing. The procedure to use this tool is discussed next.

- Click on the **Print** tool from the **File** menu; refer to Figure-30. The **Print** dialog box will be displayed; refer to Figure-31.

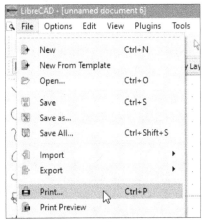

Figure-30. Print tool

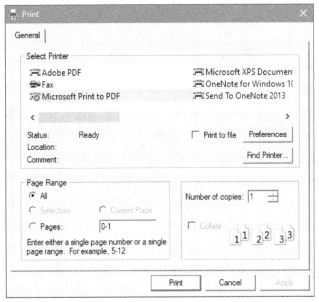

Figure-31. Print dialog box

- Select desired printer from the **Select Printer** area of the dialog box.
- Select desired page range from **Page Range** area of the dialog.
- Specify desired number of copies to be printed in the **Number of copies** edit box of the dialog box.
- After specifying desired parameters, click on the **Print** button from the dialog box to print the drawing.

Print Preview

The **Print Preview** tool is used to check the print before sending command to printer. Click on the **Print Preview** tool from the **File** menu; refer to Figure-32. The print preview of the drawing will be displayed in the print paper along with the options related to print preview; refer to Figure-33.

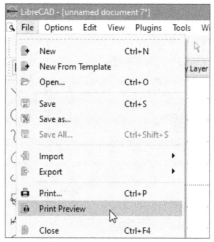

Figure-32. Print Preview tool

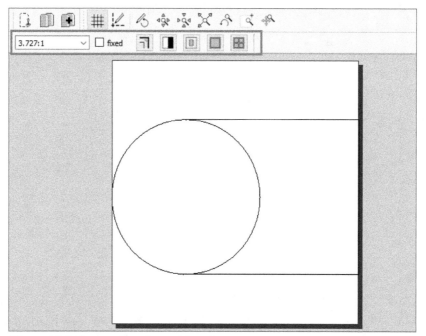

Figure-33. Print preview along with options

- Click on the **Print Scale** drop-down and select desired value to scale the print preview.
- Select the **fixed** check box to lock the current value of fixed scale selected in the **Print Scale** drop-down.
- Click on the **Apply Print Scale to line width** toggle button to apply the scale to the print according to the line width; refer to Figure-34.

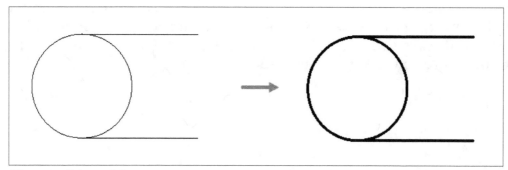

Figure-34. Print scaled

- Click on the **Toggle Black/White** mode to toggle the printing mode between black and white.
- Click on the **Center to page** toggle button to align the printing of the drawing at the center of the page; refer to Figure-35.

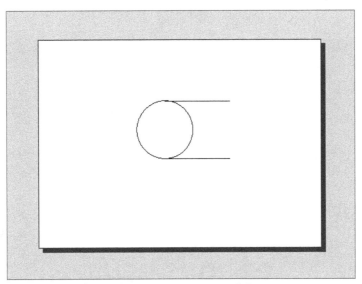

Figure-35. Print of the drawing at the center of the page

- Click on the **Fit to page** toggle button to align the printing of the drawing fit to the page; refer to Figure-36.

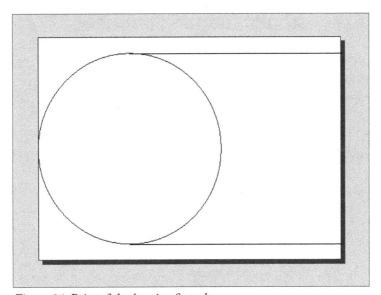

Figure-36. Print of the drawing fit to the page

- Click on the **Calculate number of pages needed to contain the drawing** toggle button to calculate the number of pages required to place the drawing to the page; refer to Figure-37.

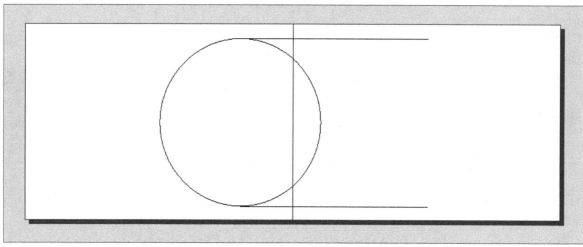

Figure-37. Number of pages calculated

- Click on the **Print Preview** tool again from the **File** menu to exit the print preview mode.

OPTIONS MENU

The **Options** menu provides tools for application preferences, current drawing preferences, widget options, device options, and reload style sheet. Various tools of **Options** menu are discussed next.

Application Preferences

In addition to the layout, LibreCAD has many preferences that will change other aspects of the appearance or behavior of the application. Different elements of the preferences can be set: Appearance, Paths, and Defaults. The procedure to use this tool is discussed next.

- Click on the **Application Preferences** tool from **Options** menu; refer to Figure-38. The **Application Preferences** dialog box will be displayed; refer to Figure-39.

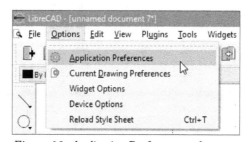

Figure-38. Application Preferences tool

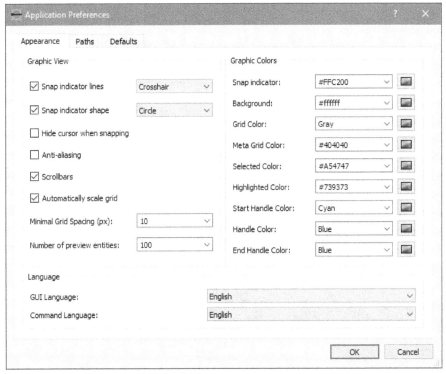

Figure–39. Application Preferences dialog box

Appearance tab

The **Appearance** tab is open by default. This tab allows the user to change the look and behavior of LibreCAD.

- The options in the **Graphic View** area includes options for the snap indicator style, shape, scroll-bars, and grid. Select **Snap indicator lines** check box to select the style for snap indicator lines viz. Crosshair, Crosshair 2, Spiderweb, and Isometric; refer to Figure-40. Select **Snap indicator shape** check box to select the style for snap indicator shape viz. Circle, Point, and Square; refer to Figure-41. Select **Anti-aliasing** check box to reduce jagged edges of diagonal lines, circles, etc.

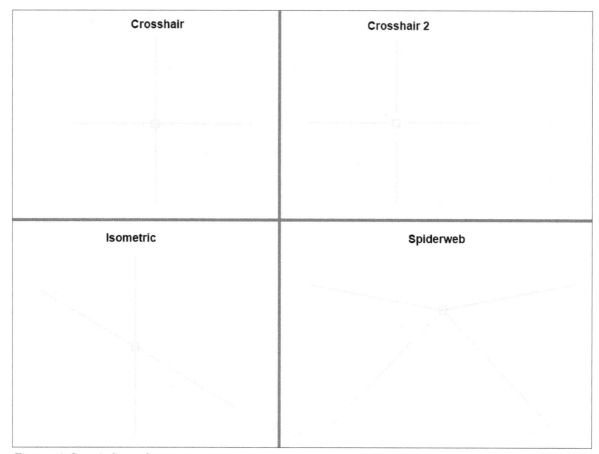

Figure-40. Snap indicator lines

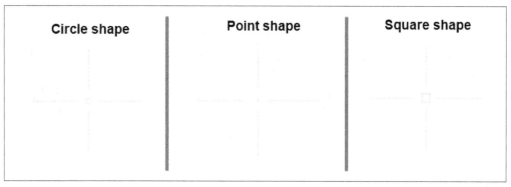

Figure-41. Snap indicator shapes

- The options in the **Graphic Colors** area allow custom colors to be selected for the snap indicator, drawing background, grid, and other indicators (selections, highlighted items, and Handles). Users can select predefined colors from the drop-down menu or select their own color from the color selector.
- The options in the **Language** area allow the user to select the language used in the GUI and command line.

Paths

- Click on the **Paths** tab of the dialog box. The options will be displayed as shown in the Figure-42.

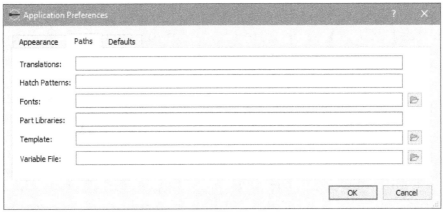

Figure–42. Paths tab of Application Preferences dialog box

- The options in this tab allow users to specify the directory paths to additional resources like Translations, Hatch Patterns, Fonts, Parts libraries, Templates, and a Variable file. These paths do not override the default paths, but are connected so the default resources are still available.
- The **Translations** field of this tab is used to specify language files for the GUI and/or command languages.
- Specify path for user defined hatch patterns, fonts, and part libraries in the **Hatch Patterns**, **Fonts**, and **Part Libraries** fields, respectively in the dialog box.
- In the **Template** field, specify the full path and filename of a user-defined drawing template to be loaded when launching the application or starting a new drawing.
- Specify the path for custom variable files in the **Variable File** field.

Defaults tab

- Click on the **Defaults** tab of the dialog box. The options will be displayed as shown in Figure-43. The **Defaults** tab allow users to specify application defaults.

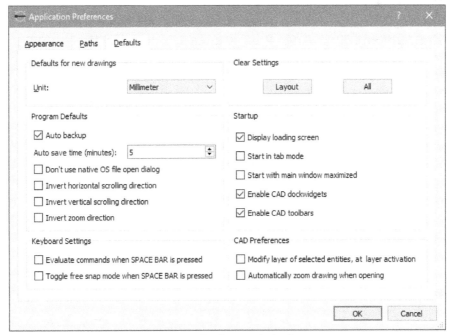

Figure–43. Defaults tab of Application Preferences dialog box

- In the **Defaults for new drawings** area of the tab, select desired unit from **Unit** drop-down to define the default unit of measurement for all new drawings.

- In the **Program Defaults** area of the tab, select **Auto backup** check box to create a backup when closing the file. Specify desired time in minutes in the **Auto save time (minutes)** edit box to perform an automatic saving of the open files. Select **Don't use native OS file open dialog** check box to display LibreCAD's file open dialog when opening files. Select **Invert horizontal scrolling direction** check box to invert scrolling direction when using mouse wheel with **Shift** key. Select **Invert vertical scrolling direction** check box to invert scrolling direction when using mouse wheel with **Ctrl** key. Select **Invert zoom direction** check box to invert zoom direction when using mouse wheel.

- In the **Clear Settings** area of the tab, click on the **Layout** button to reset the application window layout to the default configuration. Click on the **All** button to reset the application to the default configuration. Window layout, color settings, custom menus, toolbars, etc. are all reset.

- In the **Startup** area of the tab, select **Display loading screen** check box to display the LibreCAD's loading screen while launching the application. Select **Start in tab mode** check box to start LibreCAD with the tab of the drawing window. Select **Start with main window maximized** check box to start LibreCAD with the application window fullscreen. Select **Enable CAD dockwidgets** check box to show the drawing tools in the widget menu. Select **Enable CAD toolbars** check box to show drawing tools in the toolbar menu.

- In **CAD Preferences** area of the tab, select **Modify layer of selected entities, at layer activation** check box to assign the layer to the entities by selecting the entities and then selecting the destination layer from **Layer List** dialog box.

Current Drawing Preferences

The **Current Drawing Preferences** tool allow users to override the application defaults on a drawing by drawing basis. It also allow users to define specifics for the drawing's output, format, and other drawing specific configuration. The procedure to use this tool is discussed next.

- Click on the **Current Drawing Preferences** tool from the **Options** menu; refer to Figure-44. The **Drawing Preferences** dialog box will be displayed; refer to Figure-45.

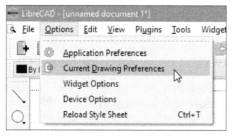

Figure-44. Current Drawing Preferences tool

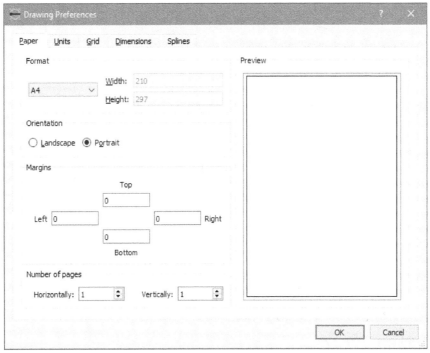

Figure–45. Drawing Preferences dialog box

Paper tab

The **Paper** tab is active by default. The **Paper** tab is used to define the size, orientation, and margins of the page used when generating output. The output can be as a physical printed page or on electronic form such as a PDF.

- In the **Format** area of the tab, select paper size from the drop-down which includes ISO, ANSI, and other sizes. Custom sizes can also be selected by choosing **Custom** option from the drop-down and specifying the paper width and height in the **Width** and **Height** edit boxes, respectively.
- In the **Orientation** area of the tab, select **Landscape** radio button for the long edge horizontal paper size and select **Portrait** radio button for the long edge vertical paper size.
- The **Margins** area of the tab determine the printable area of a page. Specify desired values in the **Top**, **Left**, **Right**, and **Bottom** edit boxes to specify margins at the edges of the page.
- In the **Number of pages** area of the tab, set values for horizontal and vertical number of pages in the **Horizontally** and **Vertically** edit boxes, respectively.

Units tab

- Click on the **Units** tab of the dialog box. The options will be displayed as shown in Figure-46. The options in **Units** tab allow users to set the main drawing unit to the preferred unit of measure and the format of linear and angular dimensions.

Figure-46. Units tab of the Drawing Preferences dialog box

- In the **Main Unit** area of the tab, select desired unit from the **Main drawing unit** drop-down.
- In the **Length** and **Angle** area of the tab, select desired format and precision from their respective drop-downs.

Grid tab

- Click on the **Grid** tab of the dialog box. The options will be displayed as shown in Figure-47. The Grid provides an evenly spaced guides to assist with placing entities.

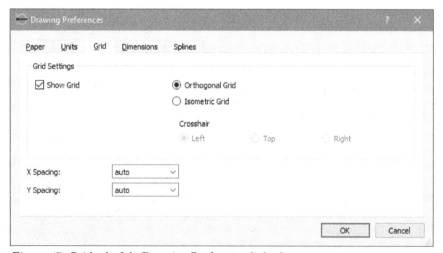

Figure-47. Grid tab of the Drawing Preferences dialog box

- In the **Grid Settings** area of the tab, select or deselect the **Show Grid** check box to toggle the grid markers between visible or not visible, respectively. Select **Orthogonal Grid** radio button to place the grid at right angles to the X and Y axis. Select **Isometric Grid** radio button to place the grid at 120^0 intervals for guiding isometric drawings. Toggle the orientation of crosshair (Left, Top, or Right) when selecting the **Isometric Grid** radio button.
- Select desired option from **X-Spacing** and **Y-Spacing** drop-downs to set the frequency of the grid markers.

Dimensions tab

- Click on the **Dimensions** tab of the dialog box. The options will be displayed as shown in Figure-48. The dimension preferences affect how dimensions appear on the drawings. The settings of the dimension includes General scale, Text size & position, Extension lines, Dimension lines, arrows and ticks, and settings for linear and angular dimensions.

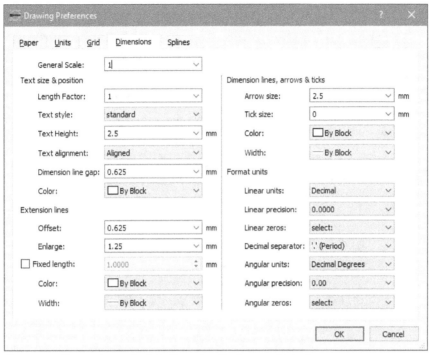

Figure-48. Dimensions tab of the Drawing Preferences dialog box

- Specify desired value in the **General Scale** edit box to adjust the size of the text and arrows.
- In the **Text size & position** area of the tab, specify desired value in the **Length Factor** edit box to specify the length of the dimension. Select desired option from **Text style** drop-down to set the font used for dimension text. Specify desired value in the **Text Height** to set the height of text. Select desired option from **Text alignment** drop-down to align the text parallel and offset to the dimension line or horizontal centered on the dimension line. Specify desired value in the **Dimension line gap** edit box to set the space between the dimension line and the dimension text. Select desired color from **Color** drop-down to set the color for the dimension lines and text.
- In the **Extension lines** area of the tab, specify desired offset value in the **Offset** edit box to specify gap between entity and dimension extension line. Specify desired value in the **Enlarge** edit box to specify length of extension line beyond dimension line. Select **Fixed length** check box and specify desired value in the edit box to fix the length of extension line measured from the dimension line towards the dimensioned entity. Select desired color and width of the line from **Color** and **Width** drop-downs, respectively.
- In the **Dimension lines, arrows & ticks** area of the tab, specify desired length of dimension arrow in the **Arrow size** edit box. Specify desired value for the tick size in the **Tick size** edit box. Select desired color and width of tick line from **Color** and **Width** drop-downs, respectively.

- In the **Format units** area of the tab, select desired units, precisions, and zeros for the linear dimensions from **Linear units**, **Linear precision**, and **Linear zeros** drop-downs, respectively. Select desired decimal separator to a period [.] or comma [,] from **Decimal separator** drop-down. Select desired units, precisions, and zeros for the angular dimensions from **Angular units**, **Angular precision**, and **Angular zeros** drop-downs, respectively.

Splines tab

- Click on the **Splines** tab of the dialog box. The option will be displayed as shown in Figure-49. Specify desired value in the **Number of line segments per spline patch** edit box which affects the smoothness of a spline. The greater the value, the smoother the spline will be drawn.

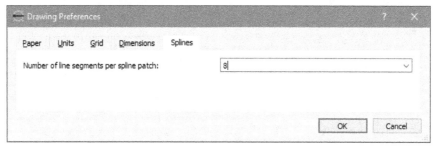

Figure-49. Splines tab of Drawing Preferences dialog box

- After specifying desired parameters, click on the **OK** button from the dialog box. The dialog box will be closed.

Widget Options

The **Widget Options** tool is used to modify the appearance of LibreCAD's border and title bar, icons, and the statusbar. The procedure to use this tool is discussed next.

- Click on the **Widget Options** tool from the **Options** menu; refer to Figure-50. The **Widget Options** dialog box will be displayed; refer to Figure-51.

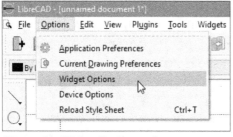

Figure-50. Widget Options tool

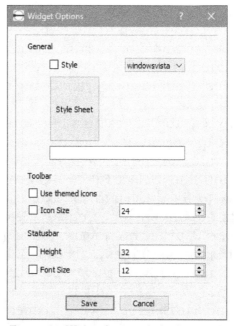

Figure-51. Widget Options dialog box

- In the **General** area of the dialog box, you can customize the LibreCAD's application window like borders, buttons, etc. and understand the use of custom style sheets. Select desired option from the drop-down to choose a style. The style list is dependent on the operating system and the version of Qt used to build LibreCAD. Note that the Qt Style Sheets is a powerful mechanism that allows you to customize the appearance of widgets. Click on the **Style Sheet** button, the **Open** dialog box will be displayed; refer to Figure-52. Select desired Qt style sheet and click on the **Open** button. The path of the Qt style sheet will be displayed in the box just below the **Style Sheet** button.

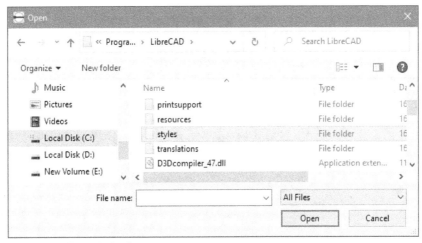

Figure-52. Open dialog box

- In the **Toolbar** area of the dialog box, select **Use themed icons** check box and change the theme of toolbar's icons. Select the **Icon Size** check box and specify the size of the icons in the **Icon Size** edit box.
- In the **Status bar** area of the dialog box, select the **Height** check box and change the height of the status bar in the **Height** edit box and select the **Font Size** check box and change the size of the font used in the **Font Size** edit box.

Device Options

The **Device Options** tool allows you to select input device: mouse, tablet, trackpad or touchscreen. Click on the **Device Options** tool from the **Options** menu; refer to Figure-53. The **Device Options** dialog box will be displayed; refer to Figure-54. Select desired input device from the drop-down in the dialog box and click on the **Set** button. Click on the **Close** button to close the dialog box.

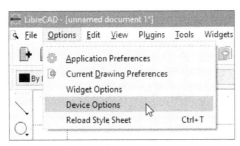

Figure-53. Device Options tool

Figure-54. Device Options dialog box

Reload Style Sheet

The **Reload Style Sheet** tool in the **Options** menu is used to reset the style sheets; refer to Figure-55.

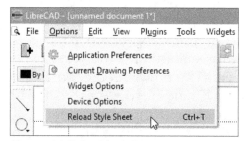

Figure-55. Reload Style Sheet tool

EDIT MENU

The **Edit** menu provides tools for selection pointer, undo, redo, cut, copy, paste, and so on; refer to Figure-56. Various tools of this menu are discussed next.

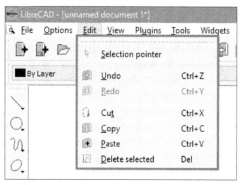

Figure-56. Edit menu

Selection Pointer

The **Selection Pointer** tool cancels the current operation and enables object selection mode.

Undo

The **Undo** tool is used to reverse the previous operations, sequentially.

Redo

The **Redo** option is used to reverse the previously reversed operations, sequentially.

Cut

The **Cut** tool is used to remove the selected entity (or entities) and places it in temporary memory, e.g. "clipboard" for later recall.

Copy

The **Copy** tool is used to create a copy of the selected entity (or entities) in temporary memory to be recalled.

Paste

The **Paste** tool is used to recall the entity (or entities) from temporary memory and place it at a location defined by a reference point.

Delete Selected

The **Delete Selected** tool is used to remove the selected entity (or entities) from the current drawing.

VIEW MENU

The **View** menu provides tools to change the view of current drawing area; refer to Figure-57. Various tools in this menu are discussed next.

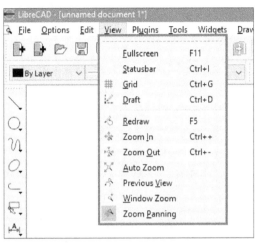

Figure-57. View menu

Fullscreen

The **Fullscreen** tool is used to hide the application title bar and toggles LibreCAD to use the entire display.

Statusbar

Select or deselect the **Statusbar** tool to toggle the visibility of the status bar at the bottom of the application window.

Grid

Select or deselect the **Grid** tool to toggle the visibility of the grid.

Draft

Select or deselect the **Draft** tool to toggle ON/OFF draft mode. Draft mode makes hatches invisible and only displays bounding boxes for images and text. Drawings will display faster, particularly on slow computers.

Redraw

The **Redraw** tool refreshes the view of the current drawing.

Zoom In

The **Zoom In** tool is used to increase the view of drawing by 25% increments.

Zoom Out

The **Zoom Out** tool is used to decrease the view of drawing by 20% increments.

Auto Zoom

The **Auto Zoom** tool is used to resize the view of the drawing to fill the drawing window.

Previous View

The **Previous View** tool is used to revert to previous zoom level of the drawing.

Window Zoom

The **Window Zoom** tool is used to increase the view of the selected area to fill the drawing window.

Zoom Panning

The **Zoom Panning** tool is used to move the view of the drawing in the window.

PLUGINS

The tools in **Plugins** menu are used to apply additional functions in the software like gear plug-in, align plug-in, and so on; refer to Figure-58. Various tools in this menu are discussed next.

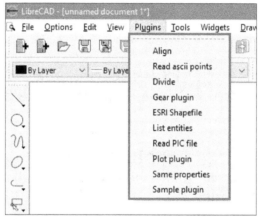

Figure-58. Plugins menu

Align

The **Align** tool is used to align selected entities to a reference by defining the final positions of 2 initial points. The procedure to use this tool is discussed next.

- Click on the **Align** tool from **Plugins** menu. You are asked to select the object to be aligned.
- Select the object which you want to align and press **Enter** from the keyboard; refer to Figure-59. You are asked to select the first base point.

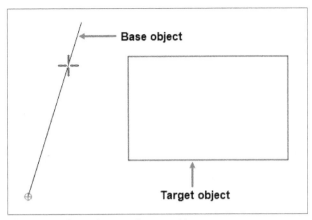

Figure-59. Base object and target object

- Select the first base point on the base object you want to align. You are asked to select the first target point.
- Select the first target point on the target object to align the first base point. You are asked to select the second base point.
- Select the second base point on the base object. You are asked to select the second target point.
- Select the second target point on the target object. The base object will be aligned to the target object; refer to Figure-60.

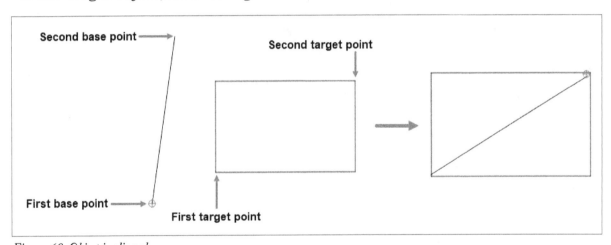

Figure-60. Object is aligned

Read ascii points

The **Read ascii points** tool is used to read points from a text file. Each line of the file is a point defined by an ID, X coordinate, Y coordinate, Z coordinate and an optional code. Each field can be separated by a comma, a tab, or a space. Click on the **Read ascii points** tool from the **Plugins** menu. The **Read ascii points** dialog box will be displayed; refer to Figure-61.

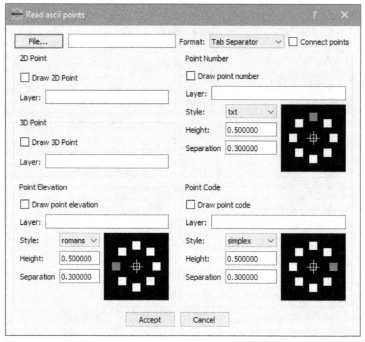

Figure-61. Read ascii points dialog box

Specify desired parameters in the dialog box and click on the **Accept** button.

Divide

The **Divide** tool is used to divide a line or a circle into n sections. A tick can be located at the limit of each section to show each limit. The size of this tick can be defined as percentage of the segment length. The line or the circle can be broken at the limit of each section using **Divide** tool. The procedure to use this tool is discussed next.

• Click on the **Divide** tool from the **Plugins** menu. You are asked to select a line, circle, or arc.

• Select desired line, circle, or arc which you want to divide and press **Enter** from the keyboard; refer to Figure-62. The **Divide** dialog box will be displayed; refer to Figure-63.

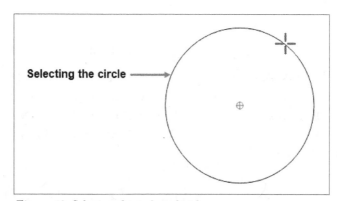

Figure-62. Selecting the circle to divide

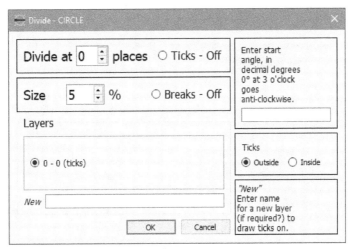

Figure-63. Divide dialog box

- Specify desired number of segments into which you want to divide the circle in the **Divide at places** edit box.
- Select the **Ticks** toggle radio button to visible or invisible the ticks on the objects.
- Specify desired value in the **Size** edit box to define the size of the tick.
- Select the **Breaks** toggle radio button to break the objects.
- Enter desired name in the **New** edit box for a new layer to draw ticks on.
- Select **Outside** radio button or **Inside** radio button from **Ticks** area of the dialog box to place the ticks outside or inside the circle, respectively.
- After specifying desired parameters in the dialog box, click on the **OK** button. The circle will be divided; refer to Figure-64.

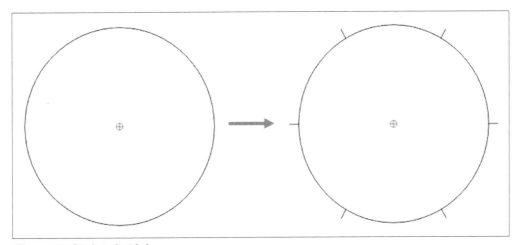

Figure-64. Circle is divided

Gear plugin

The **Gear plugin** is used to draw gear by selecting the center of gear and defining parameters such as number of teeth, modulus, and so on. The procedure to use this tool is discussed next.

- Click on the **Gear plugin** tool from **Plugins** menu. You are asked to specify the center of gear.
- Click in the Drawing Window to specify the center of gear. The **Draw a gear** dialog box will be displayed; refer to Figure-65.

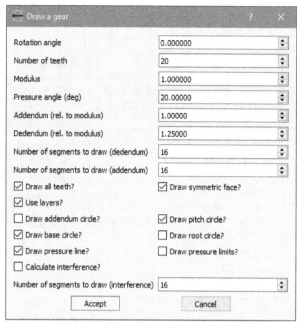

Figure-65. Draw a gear dialog box

- Specify desired parameters in the dialog box and click on the **Accept** button. The gear will be created; refer to Figure-66. The gear design illustration will be displayed in the Figure-67.

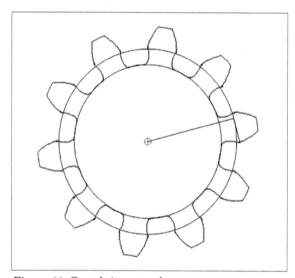

Figure-66. Gear design created

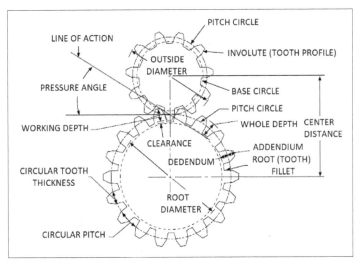

Figure-67. Gear design illustration

ESRI Shapefile

The **ESRI Shapefile** tool is used to import GIS geospatial vector data shapefile (i.e. maps). Note that the import process will lock LibreCAD until it is complete and large files can be very time consuming. The procedure to use this tool is discussed next.

- Click on the **ESRI Shapefile** tool from **Plugins** menu. The **Import ESRI Shapefile** dialog box will be displayed; refer to Figure-68.

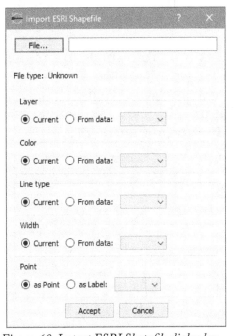

Figure-68. Import ESRI Shapefile dialog box

- Click on the **File** button from the dialog box, the **Select file** dialog box will be displayed; refer to Figure-69.

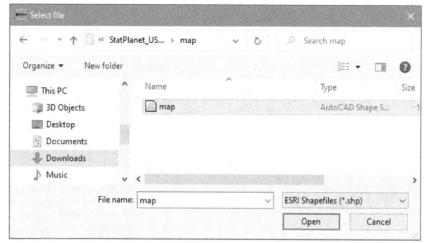

Figure-69. Select file dialog box

- Select desired file which you want to open and click on the **Open** button. The path of the file will be displayed in the edit box available next to the **File** button.
- Specify desired parameters in the dialog box and click on the **Accept** button. The file will be opened.

List entities

The **List entities** tool is used to list the selected entities along with their properties such as ID, layer, color, line type, line thickness, coordinates, and so on. The procedure to use this tool is discussed next.

- Click on the **List entities** tool from the **Plugins** menu. You are asked to select the object(s).
- Select the object(s) for which you want to list the properties and press **Enter** from the keyboard; refer to Figure-70. The **List entities** dialog box will be displayed with the properties listed; refer to Figure-71.

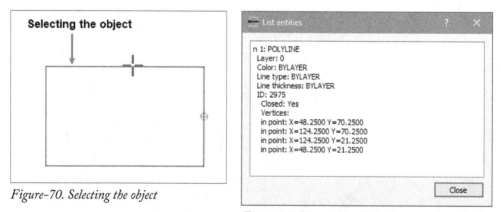

Figure-70. Selecting the object

Figure-71. List entities dialog box

- Click on the **Close** button to close the dialog box.

Read PIC file

The **Read PIC file** tool is used to import Pic graphics language diagrams. The procedure to use this tool is discussed next.

• Click on the **Read PIC file** tool from **Plugins** menu. The **LibreCAD** dialog box will be displayed; refer to Figure-72.

Figure-72. LibreCAD dialog box

• Click on the **File** button from the dialog box. The **Select file** dialog box will be displayed; refer to Figure-73.

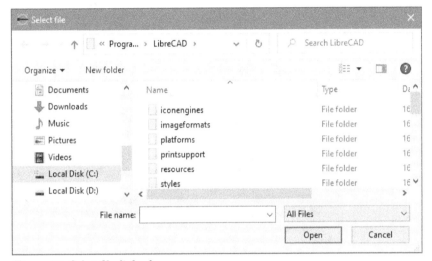

Figure-73. Select file dialog box

• Select desired file from the dialog box and click on the **Open** button. The path of selected file will be displayed in the box available next to the **File** button.
• Specify desired scale value in the **Scale** edit box to scale the file and click on the **Accept** button. The file will be opened.

Plot plugin

The **Plot Plugin** tool is used to plot a mathematical function or a parametric function using the drawing coordinate system. The formula, start value, end value, and step value are required. The plot can be lines, a polyline, or a spline. The procedure to use this tool is discussed next.

• Click on the **Plot plugin** tool from **Plugins** menu. The **Plot equation** dialog box will be displayed; refer to Figure-74.

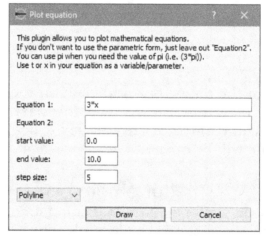

Figure-74. Plot equation dialog box

- Select desired entity, i.e., Line Segments, Polyline, or SplinePoints from the drop-down which you want to create.
- Specify desired Equation value in the **Equation 1** and **Equation 2** edit boxes.
- Specify desired start and end value of entity in the **start value** and **end value** edit boxes.
- Specify desired value in the **step size** edit box to create the entity; refer to Figure-75.

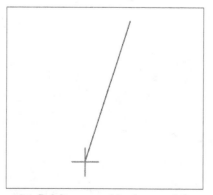

Figure-75. Plot equation dialog box after specifying parameters

- After specifying desired parameters in the dialog box, click on the **Draw** button. The entity will be created; refer to Figure-76.

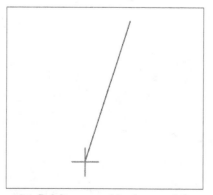

Figure-76. Polyline created by the equation

Same properties

The **Same properties** tool is used to apply the properties of a reference entity to selected entities. The modified properties are layer, color, line type, and line thickness. The procedure to use this tool is discussed next.

- Click on the **Same properties** tool from **Plugins** menu. You are asked to select the original entity.
- Select the original entity of which you want to apply the properties. You are asked to select the target entity to change the properties of entity.
- Select the target entity to which you want to apply the properties and press **Enter** from the keyboard; refer to Figure-77. The properties of the original entity will be applied to the target entity; refer to Figure-78.

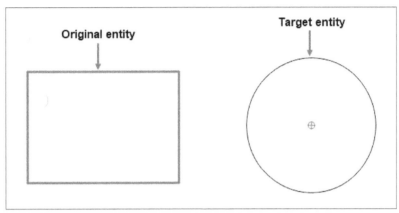

Figure-77. Selecting the entities

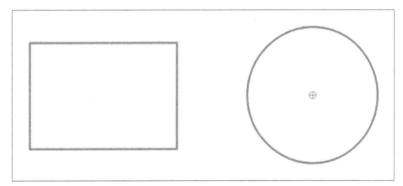

Figure-78. The properties are applied

Sample Plugin

The **Sample Plugin** tool is used to draw a line by specifying the X and Y coordinates of end points. The procedure to use this tool is discussed next.

- Click on the **Sample Plugin** tool from **Plugins** menu. The **Draw line** dialog box will be displayed; refer to Figure-79.

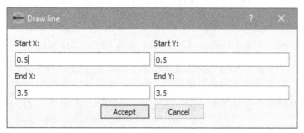

Figure-79. Draw line dialog box

- Specify desired value for the X and Y coordinates in the **Start X**, **Start Y**, **End X**, and **End Y** edit boxes of the dialog box.
- After specifying desired parameters, click on the **Accept** button from the dialog box. The line will be created; refer to Figure-80.

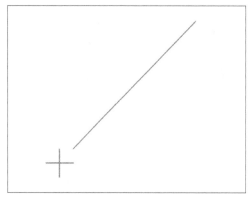

Figure-80. Line created by specifying coordinates value

WIDGETS

The **Widgets** menu provides the tools for dock areas, dock widgets, toolbars, and so on. refer to Figure-81. Various tools in this menu are discussed next.

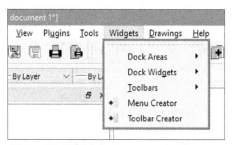

Figure-81. Widgets menu

Dock Areas

The tools in the **Dock Areas** cascading menu are used to toggle the visibility of widgets in the left, right, top, or bottom side and/or floating Dock Widgets; refer to Figure-82.

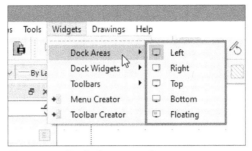

Figure-82. Dock Areas cascading menu

Dock Widgets

The **Dock Widgets** are small movable windows that serve two purposes: quick access to the drawing tools, and access to additional features which are not available in menus including block list, command line, layer list, library browser, and pen wizard; refer to Figure-83.

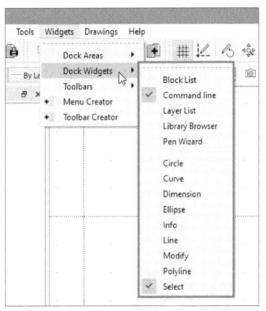

Figure-83. Dock Widgets cascading menu

Block List

The **Block List** tool in **Dock Widgets** cascading menu provides the functions to manage blocks and a list of blocks that are active in the drawing. The procedure to use this tool will be discussed later.

Command line

The **Command Line** tool is used to draw by using keyboard commands. The procedure to use this tool is discussed next.

- Click on the **Command line** tool from **Dock Widgets** cascading menu. The **Command line** widget will be displayed; refer to Figure-84.

Figure-84. Command line widget

- Specify the keycode of desired command in the **Command** edit box. The keycode will be accepted and the command will be activated; refer to Figure-85.

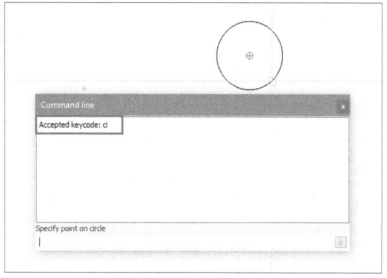

Figure-85. Command activated

Layer List

The **Layer List** tool is used to provide a list of layers in the current drawing and the functions required to manage them. The procedure to use this tool is discussed next.

- Click on the **Layer List** tool from **Dock Widgets** cascading menu. The **Layer List** widget will be displayed; refer to Figure-86.

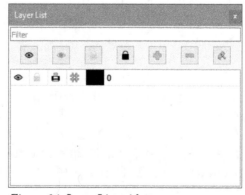

Figure-86. Layer List widget

- Click on the **Add a layer** button from the widget. The **Layer Settings** dialog box will be displayed; refer to Figure-87.

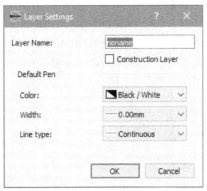

Figure-87. Layer Settings dialog box

- Specify desired name of the new layer in the **Layer Name** edit box.
- Select **Construction Layer** check box to create the layer for geometric construction.
- Select desired option from the **Default Pen** area of the dialog box. Select desired color of the layer from **Color** drop-down. Select desired width of the layer from **Width** drop-down. Select desired line type of the layer from **Line type** drop-down. Note that options specified here will be applied to all the objects created on current layers.
- After specifying desired parameters, click on the **OK** button from the dialog box. The new layer will be created; refer to Figure-88 and Figure-89.

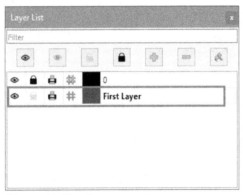

Figure-88. New layer created

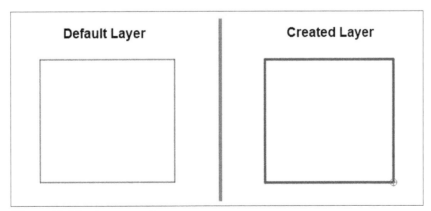

Figure-89. New layer created

- Click on the **Show all layers** button from the dialog box to show all the hidden layers.
- Click on the **Hide all layers** button to hide all the visible layers.

- Click on the **Unlock all layers** ▦ button to unlock all the locked layers.
- Click on the **Lock all layers** 🔒 button to lock all the unlocked layers.
- Click on the **Remove layer** ▬ button to remove all the layers present in the list.
- Click on the **Modify layer attributes/rename** 🖉 button to modify the attributes of layer. The **Layer Settings** dialog box will be displayed as discussed earlier.
- Click on the toggle print 🖨 button to enable or disable the printing of selected layer.
- Click on the toggle construction ▦ button to create the construction layers which are used for reference geometry.

Library Browser

The **Library Browser** tool is used to insert blocks into the current drawing available in the defined libraries. The procedure to use this tool is discussed next.

- Click on the **Library Browser** tool. The **Library Browser** dialog box will be displayed; refer to Figure-90.

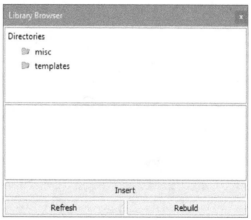

Figure-90. Library Browser dialog box

- Select desired block category from **Directories** area of the dialog box. The blocks related to that category will be displayed in the dialog box; refer to Figure-91.

Figure-91. Blocks displayed

- Select desired block from the dialog box and click on the **Insert** button. The selected block will be attached to the cursor and you will be asked to specify the insertion point.

- Click in the Drawing Window at desired location to place the block. The block will be inserted; refer to Figure-92.

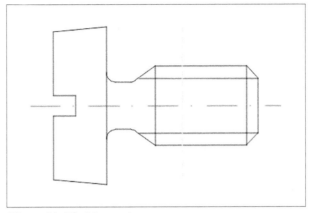

Figure-92. Block inserted

Pen Wizard

The **Pen Wizard** tool is used to create a palette of favorite colors for the drawing tools. The procedure to use this tool is discussed next.

- Click on the **Pen Wizard** tool from **Dock Widgets** cascading menu. The **Pen Wizard** dialog box will be displayed; refer to Figure-93.

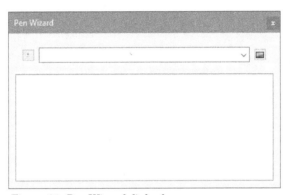

Figure-93. Pen Wizard dialog box

- Select desired color from the drop-down or from the **Select Color** dialog box displayed on clicking the ▣ button in the **Pen Wizard** dialog box; refer to Figure-94.

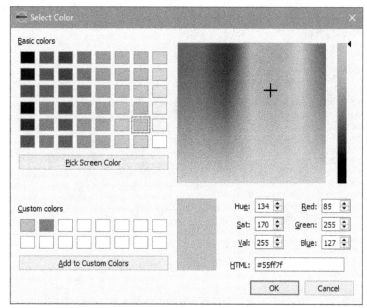

Figure-94. Select Color dialog box

- After selecting desired color, click on the **Add to favorites** button to add the selected color to the list box; refer to Figure-95.

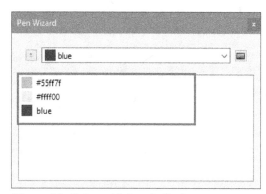

Figure-95. Colors added to the list box

- Now, right click on desired color to be applied to the object. The options will be displayed as shown in Figure-96.

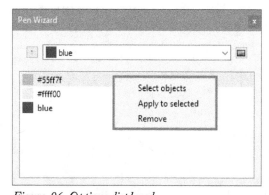

Figure-96. Options displayed

- Click on the **Select objects** option to select the objects of respective color from drawing area.
- Click on the **Apply to selected** option to apply selected color to objects selected in the drawing area.; refer to Figure-97.

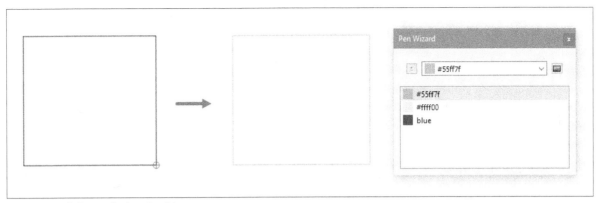

Figure-97. Color applied to the object

• Click on the **Remove** option to remove the color from the list box.

Toolbars

Toolbars provide an alternative to the menus for accessing application functions and drawing tools. Toolbars can be moved anywhere on the display and left floating, or docked to any of the four sides of the drawing window; refer to Figure-98.

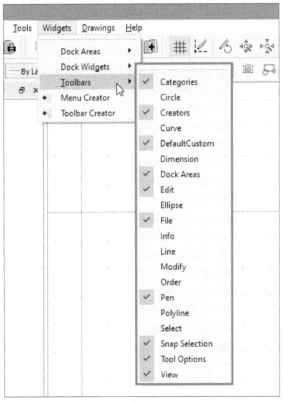

Figure-98. Toolbars cascading menu

Menu Creator

The **Menu Creator** is used to create, modify, or delete custom pop-up menus. The procedure to use this tool is discussed next.

• Click on the **Menu Creator** tool from **Widgets** menu. The **Menu Creator** dialog box will be displayed; refer to Figure-99.

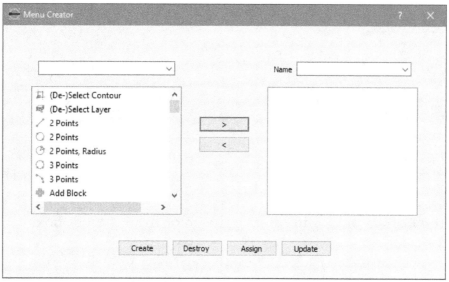

Figure-99. Menu Creator dialog box

- Specify desired name for the menu in the **Name** edit box at the right hand side.
- Select desired toolbar from the drop-down at the left hand side.
- Select desired tool from the left-hand list box which you want to add to the new menu at the right-hand list box and click on the arrow ⟨ > ⟩ button. The tool will be added to the new menu; refer to Figure-100.

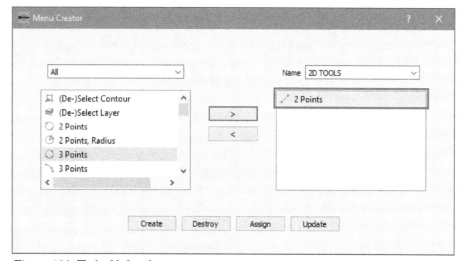

Figure-100. Tool added to the new menu

- Move & drop the tool in the right-hand list box to arrange the tool upwards or downwards.
- If you want to remove the tool from the menu, select the tool from the new menu created and click on the arrow ⟨ < ⟩ button.
- Click on the **Create** button to create the menu.
- Similarly, you can modify existing menu and then click on the **Update** button. The menu will be updated.
- If you want to delete a menu then select the menu from the **Name** drop-down and click on the **Destroy** button.
- Click on the **Assign** button from the dialog box to assign activation method for the menu. The **Menu Assigner** dialog box will be displayed; refer to Figure-101.

Figure-101. Menu Assigner dialog box

- Select desired check box(es) from the dialog box to assign a custom pop-up menu to a mouse button.
- After selecting desired check boxes, click on the **Save** button from the dialog box.
- Click on the **Close** button to close the **Menu Creator** dialog box.

Toolbar Creator

The **Toolbar Creator** is used to create, modify, or delete the custom toolbars. The procedure to use the **Toolbar Creator** is same as discussed in **Menu Creator**.

DRAWINGS MENU

The **Drawings** menu provides the tools for tab mode, window mode, layout, arrange, etc; refer to Figure-102. Various tools in this menu are discussed next.

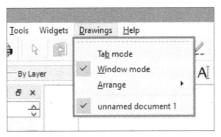

Figure-102. Drawings menu

Tab mode

Select the **Tab mode** button to display current open drawings in tabs that are the left in the application window. Tabbed windows cannot be moved, resized, or arranged within the LibreCAD application window. The selected tab is the active window; refer to Figure-103.

Window mode

Select the **Window mode** button to display each open drawing in its own independent application window. Application windows can be moved, resized, or arranged. Window mode is the default mode; refer to Figure-103.

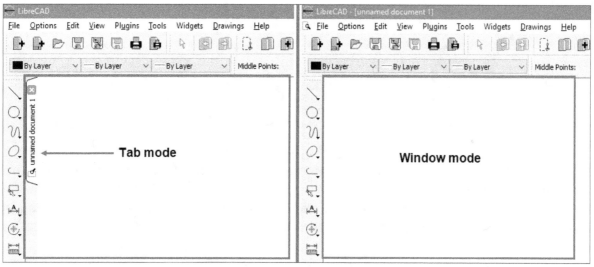

Figure-103. Tab mode and Window mode

Arrange cascading menu

The options in the **Arrange** cascading menu work only in **Window mode**. Drawing windows can be maximize, cascade, tiled, tiled vertically, or tiled horizontally; refer to Figure-104.

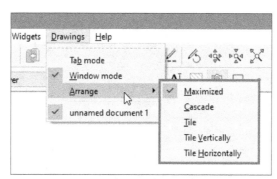

Figure-104. Arrange cascading menu

Currently opened drawings

The Currently opened drawings list shows the drawings which are currently open in the software. The item with the checked box is the active drawing.

HELP MENU

The **Help** menu provides tools to access online help, version information about the software, license details, and so on; refer to Figure-105. Various tools in this menu are discussed next.

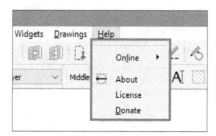

Figure-105. Help menu

Online cascading menu

The **Online** cascading menu provides links to online resources; 'Wiki', 'User's Manual', 'Command', 'Style Sheets', 'Widgets', 'Forum' and 'Release Information'; refer to Figure-106.

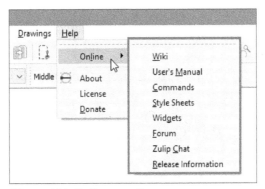

Figure-106. Online cascading menu

About

The **About** option displays information about the current version of LibreCAD and web links of the 'Contributors', 'License' and 'The Code' repository.

License

The **License** option displays the license text.

SELF ASSESSMENT

Q1. What is the use of Read ascii points tool?

Q2. What is the use of Gear plugin tool?

Q3. Which of the following tools is used to save the file with different name?

a) Save
b) Save All
c) Save As
d) Save with

Q4. Which of the following tabs in Application Preferences dialog box allow the user to change the look and behavior of LibreCAD?

a) Appearance
b) Paths
c) Defaults
d) Graphics

Q5. Which of the following tools is used to plot a mathematical function or a parametric function using the drawing coordinate system?

a) Sample plugin
b) Read PIC file
c) Plot plugin
d) None of the above

Q6. The tools in the _____ cascading menu are used to export current drawing file into various different formats.

Q7. The _____ tool is used to create a palette of favorite colors for the drawing tools.

Q8. The _____ menu provides tools for application preferences, current drawing preferences, widget options, device options, and reload style sheet.

Q9. The **Fullscreen** tool is used to hide the application title bar and toggles LibreCAD to use the entire display. (True/False)

Q10. The **Auto Zoom** tool is used to increase the view of the selected area to fill the drawing window. (True/False)

Q11. The **ESRI Shapefile** tool is used to import GIS geospatial vector data shapefile (i.e. maps). (True/False)

Ans: **3.** (c) **4.** (a) **5.** (c) **6.** Export **7.** Pen Wizard **8.** Options **9.** True **10.** False **11.** True

Chapter 2

Sketching

Topics Covered

The major topics covered in this chapter are:

- *Introduction*
- *Starting Sketch*
- *Sketch Creation Tools*
- *Sketch Editing Tools*
- *Sketcher Constraints*
- *Sketcher Tools*
- *Sketcher B-Spline Tools*
- *Sketching Practical and Practice*

INTRODUCTION

In Engineering, sketches are based on real dimensions of real world objects. These sketches work as building blocks for various 2D operations. In this chapter, we will be working with sketch entities like; Line, Circle, Arc, Polygon, Ellipse, and so on.

In this chapter, we will be working on **Drawing Tools**. We will learn about various tools of toolbar used in 2D sketching.

LINE CREATION TOOLS

Line creation tools are displayed in the **Lines** drop-down; refer to Figure-1. You can also select these tools from **Line** cascading menu of **Tools** menu; refer to Figure-2 or **Line** widget; refer to Figure-3 or **Line** toolbar; refer to Figure-4. These tools are discussed next.

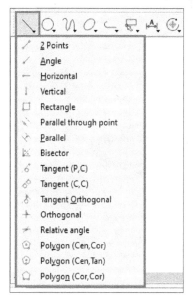

Figure-1. Lines drop down

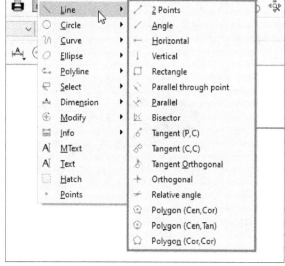

Figure-2. Line creation tools from Tools menu

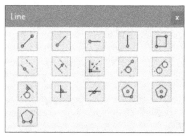

Figure-3. Line widget

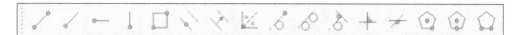

Figure-4. Line toolbar

2 Points

Toolbar: Line > 2 Points
Widget: Line > 2 Points
Drop-Down: Lines > 2 Points
Menu: Tools > Line > 2 Points

The **2 Points** tool is used to draw a line between two assigned points. The procedure to use this tool is discussed next.

- Click on the **2 Points** tool. The **Tool Options** will be displayed in the toolbar; refer to Figure-5 and you will be asked to specify the first point.

Figure-5. 2 Points Tool Options

- Click in the Drawing Window to specify the first point. You will be asked to specify the next point.
- Move the cursor away and click in the Drawing Window to specify the next point. The line will be created; refer to Figure-6.

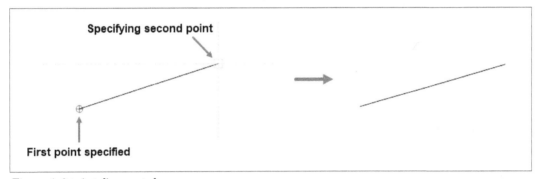

Figure-6. 2 points line created

- Click on the **Close** button to form a closed contour from lines drawn; refer to Figure-7.

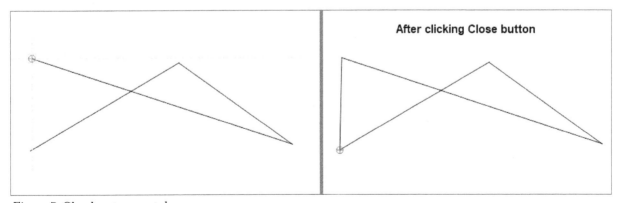

Figure-7. Closed contour created

- Press **ESC** button or click **RMB** to exit the tool.

Angle

Toolbar: Line > Angle
Widget: Line > Angle
Drop-Down: Lines > Angle
Menu: Tools > Line > Angle

The **Angle** tool is used to draw a line from an assigned point defining the start, middle, or end of the line or with an assigned length and angle. The procedure to use this tool is discussed next.

- Click on the **Angle** tool. The **Tool Options** will be displayed; refer to Figure-8 and a line will be attached to the cursor.

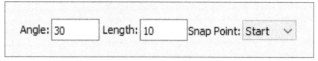

Figure-8. Angle Tool Options

- Specify desired angle and length of line in the **Angle** and **Length** edit boxes, respectively.
- Select desired option from **Snap Point** drop-down to create the line from start point, middle point, or end point.
- After specifying desired parameters, click in the Drawing Window. The line will be created; refer to Figure-9.

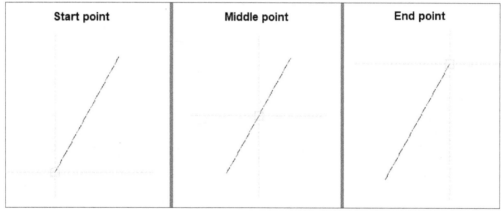

Figure-9. Line created with Snap Point options

- Press **Esc** button or click **RMB** to exit the tool.

Horizontal

Toolbar: Line > Horizontal
Widget: Line > Horizontal
Drop-Down: Lines > Horizontal
Menu: Tools > Line > Horizontal

The **Horizontal** tool is used to draw a horizontal line from an assigned point defining the start, middle, or end of the line and with an assigned length. The procedure to use this tool is discussed next.

- Click on the **Horizontal** tool. The **Tool Options** will be displayed; refer to Figure-10 and you will be asked to specify the first point.

Figure-10. Horizontal Tool Options

- Specify desired length of line in the **Length** edit box.
- Select desired option from **Snap Point** drop-down. The options in the **Snap Point** drop-down has been discussed earlier.
- After specifying desired parameters, click in the Drawing Window to place the line. The horizontal line will be created.
- Press **Esc** button or click **RMB** to exit the tool.

Vertical

Toolbar: Line > Vertical
Widget: Line > Vertical
Drop-Down: Lines > Vertical
Menu: Tools > Line > Vertical

The **Vertical** tool is used to draw a vertical line from an assigned point defining the start, middle, or end of the line and with an assigned length. The procedure to use this tool is same as discussed in the previous tool.

Rectangle

Toolbar: Line > Rectangle
Widget: Line > Rectangle
Drop-Down: Lines > Rectangle
Menu: Tools > Line > Rectangle

The **Rectangle** tool is used to draw a rectangle by assigning the points of two diagonally opposite corners. The procedure to use this tool is discussed next.

- Click on the **Rectangle** tool. You will be asked to specify the first corner point.
- Click in the Drawing Window to specify the first corner point. You will be asked to specify the second corner point.
- Move the cursor away and click in the Drawing Window. The rectangle will be created; refer to Figure-11.

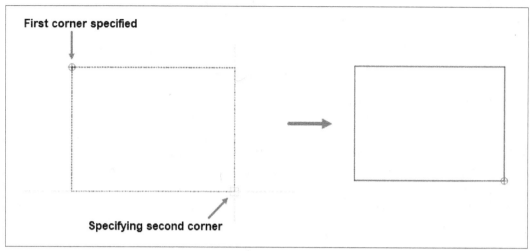

First corner specified

Specifying second corner

Figure-11. Rectangle created

• Press **Esc** button or click **RMB** to exit the tool.

Parallel through point

Toolbar: Line > Parallel through point
Widget: Line > Parallel through point
Drop-Down: Lines > Parallel through point
Menu: Tools > Line > Parallel through point

The **Parallel through point** tool is used to draw specified number of lines parallel to a selected existing line through an assigned point. The procedure to use this tool is discussed next.

• Click on the **Parallel through point** tool. The **Tool Options** will be displayed; refer to Figure-12 and you will be asked to select the object.

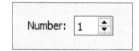

Number: 1

Figure-12. Parallel through point Tool Options

• Specify desired number of parallel line(s) to be created in the **Number** edit box.
• Select the object whose parallel line(s) you want to create.
• Move the cursor away and click at desired location to place the line(s). The parallel line(s) will be created; refer to Figure-13.

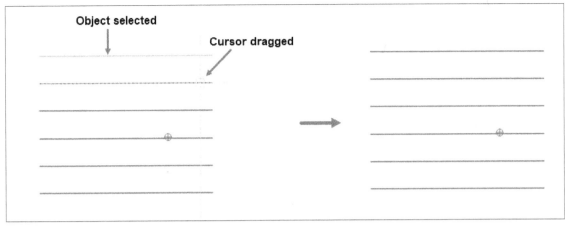

Figure-13. Parallel lines created

• Press **Esc** button or click **RMB** to exit the tool.

Parallel

Toolbar: Line > Parallel
Widget: Line > Parallel
Drop-Down: Lines > Parallel
Menu: Tools > Line > Parallel

The **Parallel** tool is used to draw specified number of lines parallel to a selected existing line with a given distance between lines. The procedure to use this tool is discussed next. The procedure to use this tool is discussed next.

• Click on the **Parallel** tool. The **Tool Options** will be displayed; refer to Figure-14 and you will be asked to select the object.

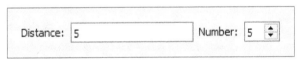

Figure-14. Parallel Tool Options

• Specify desired distance of parallel line(s) from the original object in the **Distance** edit box.
• Specify desired number of parallel line(s) to be created in the **Number** edit box.
• As you hover the cursor on the original object. The preview of parallel line(s) will be displayed.
• Click on the original object to create the line(s). The parallel line(s) will be created.
• Press **Esc** button or click **RMB** to exit the tool.

Bisector

Toolbar: Line > Bisector
Widget: Line > Bisector
Drop-Down: Lines > Bisector
Menu: Tools > Line > Bisector

The **Bisector** tool is used to draw specified number of lines bisecting two existing non-parallel lines. The procedure to use this tool is discussed next.

- Click on the **Bisector** tool. The **Tool Options** will be displayed; refer to Figure-15 and you will be asked to select the first line.

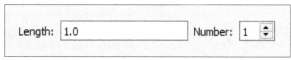

Figure-15. Bisector Tool Options

- Specify desired length of bisector in the **Length** edit box.
- Specify desired number of bisector line(s) to be created in the **Number** edit box.
- Select the first line and hover the cursor on second line. The preview of bisector line will be displayed.
- Click on the second line. The bisector line will be created; refer to Figure-16.

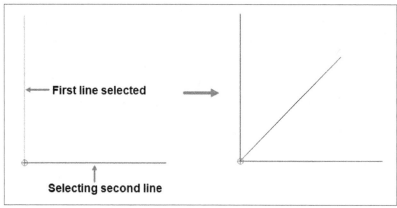

Figure-16. Bisector line created

- Press **Esc** button or click **RMB** to exit the tool.

Tangent (P,C)

Toolbar: Line > Tangent (P,C)
Widget: Line > Tangent (P,C)
Drop-Down: Lines > Tangent (P,C)
Menu: Tools > Line > Tangent (P,C)

The **Tangent (P,C)** tool is used to draw a line from an assigned point tangent to an existing circle. The procedure to use this tool is discussed next.

- Click on the **Tangent (P,C)** tool. You will be asked to specify the point.
- Click in the Drawing Window at desired location to specify the start point. You will be asked to select the circle, arc, or ellipse.
- As you hover the cursor on the circle. The preview of tangent line will be displayed.
- Click on the circle. The tangent line will be created; refer to Figure-17.

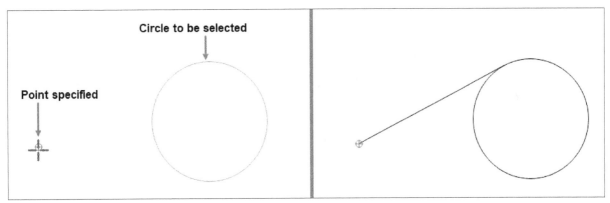

Figure-17. Tangent line created

- Press **Esc** button or click **RMB** to exit the tool.

Tangent (C,C)

Toolbar: Line > Tangent (C,C)
Widget: Line > Tangent (C,C)
Drop-Down: Lines > Tangent (C,C)
Menu: Tools > Line > Tangent (C,C)

The **Tangent (C,C)** tool is used to draw a line tangent to two existing circles. The procedure to use this tool is discussed next.

- Click on the **Tangent (C,C)** tool. You will be asked to select first circle or ellipse.
- Select the first circle in the Drawing Window. You will be asked to select the second circle or ellipse.
- As you hover the cursor on the second circle, the preview of tangent line will be displayed.
- Click on the second circle. The tangent line will be created; refer to Figure-18.

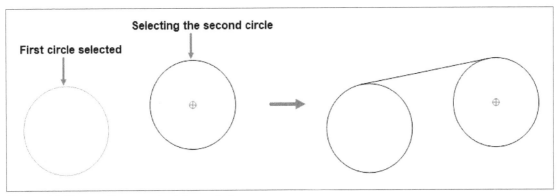

Figure-18. Tangent line created with two circles

- Press **Esc** button or click **RMB** to exit the tool.

Tangent Orthogonal

Toolbar: Line > Tangent Orthogonal
Widget: Line > Tangent Orthogonal
Drop-Down: Lines > Tangent Orthogonal
Menu: Tools > Line > Tangent Orthogonal

The **Tangent Orthogonal** tool is used to draw a line tangent to an existing circle and perpendicular to an existing line. The procedure to use this tool is discussed next.

- Click on the **Tangent Orthogonal** tool. You will be asked to select a line.
- Select desired line perpendicular to which the tangent line is to be created. You will be asked to select the circle.
- As you hover the cursor on the circle, the preview of tangent line will be displayed.
- Click on the circle. The tangent line will be created; refer to Figure-19.

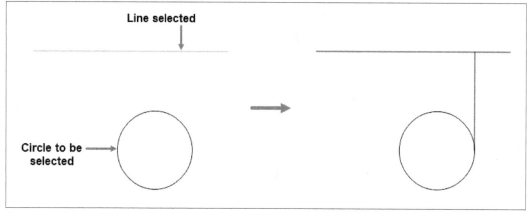

Figure-19. Tangent line created orthogonally

- Press **Esc** button or click **RMB** to exit the tool.

Orthogonal

Toolbar: Line > Orthogonal
Widget: Line > Orthogonal
Drop-Down: Lines > Orthogonal
Menu: Tools > Line > Orthogonal

The **Orthogonal** tool is used to draw a line of given length perpendicular to an existing line while placing the centre at an assigned point. The procedure to use this tool is discussed next.

- Click on the **Orthogonal** tool. The **Tool Options** will be displayed; refer to Figure-20 and you will be asked to select the base entity.

Figure-20. Orthogonal Tool Options

- Specify desired length of line to be created in the **Length** edit box.
- Select the base entity perpendicular to which the line to be created. A line perpendicular to the base entity will be attached to the cursor and you will be asked to specify the position of line.
- Click at desired location in the Drawing Window to place the line. The orthogonal line will be created; refer to Figure-21.

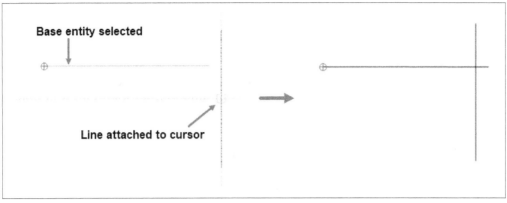

Figure-21. Orthogonal line created

- Press **Esc** button or click **RMB** to exit the tool.

Relative Angle

Toolbar: Line > Relative Angle
Widget: Line > Relative Angle
Drop-Down: Lines > Relative Angle
Menu: Tools > Line > Relative Angle

The **Relative Angle** tool is used to draw a line of specified length and at a given angle relative to an existing line placing the center of the line at an assigned point. The procedure to use this tool is discussed next.

- Click on the **Relative Angle** tool. The **Tool Options** will be displayed; refer to Figure-22 and you will be asked to select the base entity.

Figure-22. Relative Angle Tool Options

- Specify desired angle and length of line in the **Angle** and **Length** edit boxes, respectively.
- Select the base entity relative to which the angled line will be created. The angled line will be attached to cursor and you will be asked to specify the position of line.
- Click in the Drawing Window at desired location to place the angled line. The angled line will be created; refer to Figure-23.

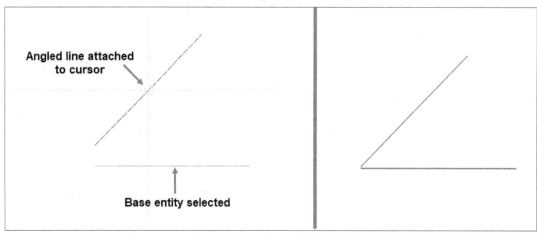

Figure-23. Angled line created

• Press **Esc** button or click **RMB** to exit the tool.

Polygon (Cen,Cor)

Toolbar: Line > Polygon (Cen,Cor)

Widget: Line > Polygon (Cen,Cor)

Drop-Down: Lines > Polygon (Cen,Cor)

Menu: Tools > Line > Polygon (Cen,Cor)

The **Polygon (Cen,Cor)** tool is used to draw a polygon of specified number of sides assigning the center point and one vertex. The procedure to use this tool is discussed next.

• Click on the **Polygon (Cen,Cor)** tool. The **Tool Options** will be displayed; refer to Figure-24 and you will be asked to specify the center of polygon.

Figure-24. Polygon Cen Cor Tool Options

• Specify desired number of edges to be created in the polygon in the **Number** edit box.
• Click in the Drawing Window at desired location to specify the center of polygon. You will be asked to specify the corner point of polygon.
• Move the cursor away and click at desired location to specify the corner point of polygon. The polygon will be created; refer to Figure-25.

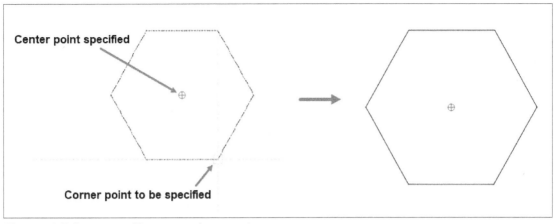

Figure-25. Polygon created by center and corner point

• Press **Esc** button or click **RMB** to exit the tool.

Polygon (Cen,Tan)

Toolbar: Line > Polygon (Cen,Tan)
Widget: Line > Polygon (Cen,Tan)
Drop-Down: Lines > Polygon (Cen,Tan)
Menu: Tools > Line > Polygon (Cen,Tan)

The **Polygon (Cen,Tan)** tool is used to draw a polygon with specified number of sides assigning the center point and point of center of one side. The procedure to use this tool is discussed next.

• Click on the **Polygon (Cen,Tan)** tool. The **Tool Options** will be displayed; refer to Figure-24 and you will be asked to specify the center point of polygon.
• Specify desired number of edges to be created in the polygon in the **Number** edit box.
• Click in the Drawing Window at desired location to specify the center point of polygon. You will be asked to specify a tangent (circumferential) point.
• Move the cursor away and click at desired location to specify the tangent for polygon. The polygon will be created; refer to Figure-26.

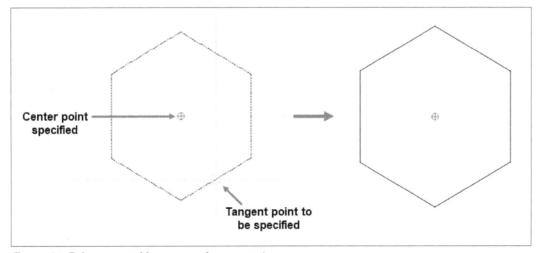

Figure-26. Polygon created by center and tangent points

• Press **Esc** button or click **RMB** to exit the tool.

Polygon (Cor,Cor)

Toolbar: Line > Polygon (Cor,Cor)
Widget: Line > Polygon (Cor,Cor)
Drop-Down: Lines > Polygon (Cor,Cor)
Menu: Tools > Line > Polygon (Cor,Cor)

The **Polygon (Cor,Cor)** tool is used to draw a polygon with specified number of sides by assigning the two points of one side. The procedure to use this tool is discussed next.

- Click on the **Polygon (Cor,Cor)** tool. The **Tool Options** will be displayed; refer to Figure-24 and you will be asked to specify the first corner point of polygon.
- Specify desired number of edges to be created in the polygon in the **Number** edit box.
- Click in the Drawing Window at desired location to specify the first corner point of polygon. You will be asked to specify the second corner point of polygon.
- Move the cursor away and click at desired location to specify the second corner point of polygon. The polygon will be created; refer to Figure-27.

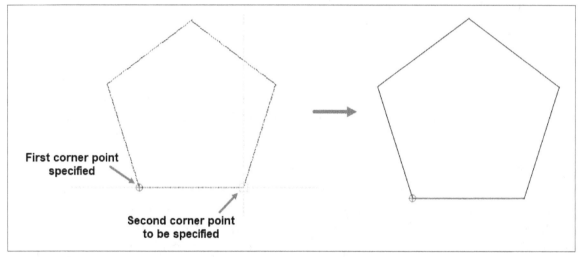

First corner point
specified

Second corner point
to be specified

Figure-27. Polygon created by using two corner points

- Press **Esc** button or click **RMB** to exit the tool.

CIRCLE CREATION TOOLS

Circle creation tools are available in the **Circles** drop-down; refer to Figure-28. You can also select these tools from **Circle** cascading menu in **Tools** menu; refer to Figure-29 or **Circle** widget; refer to Figure-30 or **Circle** toolbar; refer to Figure-31.

These tools are discussed next.

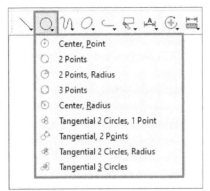

Figure-28. Circles drop down

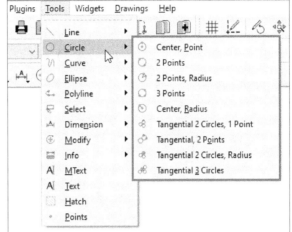

Figure-29. Circle creation tools from Tools menu

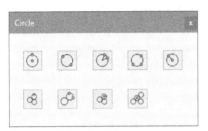

Figure-31. Circle toolbar

Figure-30. Circle widget

Creating circle using Center, Point tool

Toolbar: Circle > Center, Point
Widget: Circle > Center, Point
Drop-Down: Circles > Center, Point
Menu: Tools > Circle > Center, Point

The **Center, Point** tool is used to draw a circle of given radius by assigning a center point and a point on the circumference. The procedure to use this tool is discussed next.

- Click on the **Center, Point** tool. You will be asked to specify the center of circle.
- Click in the Drawing Window at desired location to specify the center of circle. You will be asked to specify the point on circle.
- Move the cursor away and click at desired location to specify the point on the circumference of circle. The circle will be created; refer to Figure-32.

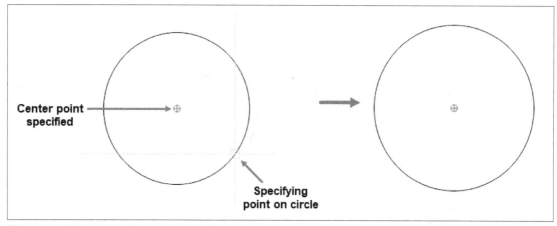

Figure-32. Circle created using center and point

- Press **Esc** button or click **RMB** to exit the tool.

Creating circle using 2 Points tool

Toolbar: Circle > 2 Points
Widget: Circle > 2 Points
Drop-Down: Circles > 2 Points
Menu: Tools > Circle > 2 Points

The **2 Points** tool is used to draw a circle of specified diameter by assigning two opposite points on the circumference. The procedure to use this tool is discussed next.

- Click on the **2 Points** tool. You will be asked to specify the first point.
- Click in the Drawing Window at desired location to specify the first point of circumference of the circle. You will be asked to specify the second point.
- Move the cursor away and click at desired location to specify second point on the circumference of circle. The circle will be created using 2 points; refer to Figure-33.

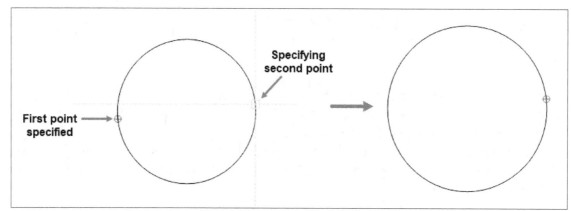

Figure-33. 2 points circle created

- Press **Esc** button or click **RMB** to exit the tool.

Creating circle using 2 Points, Radius tool

Toolbar: Circle > 2 Points, Radius
Widget: Circle > 2 Points, Radius
Drop-Down: Circles > 2 Points, Radius
Menu: Tools > Circle > 2 Points, Radius

The **2 Points, Radius** tool is used to draw a circle with two points on the circumference and with an assigned radius. The procedure to use this tool is discussed next.

• Click on the **2 Points, Radius** tool. The **Tool Options** will be displayed; refer to Figure-34 and you will be asked to specify the first point.

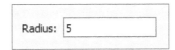

Figure-34. 2 Points Radius Tool Options

• Specify desired radius of circle to be created in the **Radius** edit box.
• Click in the Drawing Window at desired location to specify the first point on the circumference of circle. You will be asked to specify the second point.
• Move the cursor away and click at desired location near the first point to specify second point on the circumference of circle. You will be asked to specify the center of circle.
• Move the cursor away and click at desired location to specify the center of circle. The circle will be created; refer to Figure-35.

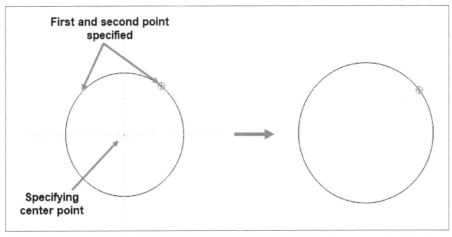

Figure-35. Circle created using 2 points and radius

• Press **Esc** button or click **RMB** to exit the tool.

Creating circle using 3 Points tool

Toolbar: Circle > 3 Points
Widget: Circle > 3 Points
Drop-Down: Circles > 3 Points
Menu: Tools > Circle > 3 Points

The **3 Points** tool is used to draw a circle by assigning three points on the circumference. The procedure to use this tool is discussed next.

- Click on the **3 Points** tool. You will be asked to specify the second point.
- Click in the Drawing Window at desired location to specify the first point on the circumference of circle. You will be asked to specify the second point.
- Move the cursor away and click at desired location to specify the second point on the circumference of circle. You will be asked to specify the third point.
- Similarly, specify third point on the circumference of the circle. The circle will be created; refer to Figure-36.

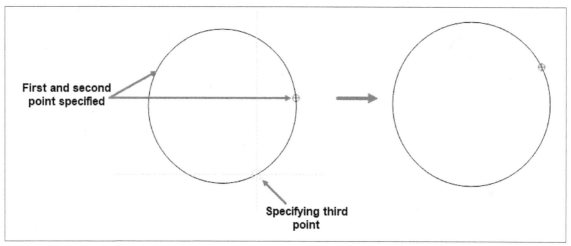

Figure-36. Circle created using 3 points

- Press **Esc** button or click **RMB** to exit the tool.

Creating circle using Center, Radius tool

Toolbar: Circle > Center, Radius
Widget: Circle > Center, Radius
Drop-Down: Circles > Center, Radius
Menu: Tools > Circle > Center, Radius

The **Center, Radius** tool is used to draw a circle with radius and center point. The procedure to use this tool is discussed next.

- Click on the **Center, Radius** tool. The **Tool Options** will be displayed; refer to Figure-37 along with the circle attached to cursor. You will be asked to specify the center of circle.

Figure-37. Radius edit box

- Specify desired radius of the circle to be created in the **Radius** edit box. The specified radius will be applied to the circle attached to the cursor.
- Click in the Drawing Window at desired location to specify the center of circle. The circle will be created; refer to Figure-38.

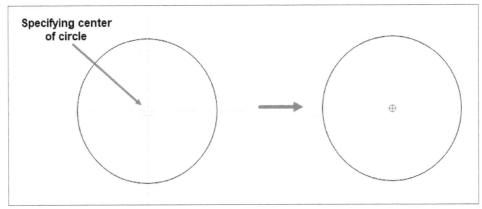

Figure-38. Circle created using center and radius

- Press **Esc** button or click **RMB** to exit the tool.

Creating circle using Tangential 2 Circles, 1 Point tool

Toolbar: Circle > Tangential 2 Circles, 1 Point
Widget: Circle > Tangential 2 Circles, 1 Point
Drop-Down: Circles > Tangential 2 Circles, 1 Point
Menu: Tools > Circle > Tangential 2 Circles, 1 Point

The **Tangential 2 Circles, 1 Point** tool is used to draw a circle tangential to two existing circles/lines/arcs and assigning a center point to establish the radius. The procedure to use this tool is discussed next.

- Click on the **Tangential 2 Circles, 1 Point** tool. You will be asked to select line, arc, or circle.
- Select desired line, arc, or circle in the Drawing Window as the first tangent entity You will be asked to select another line, arc, or circle.
- Select desired line, arc, or circle as the second tangent entity. You will be asked to specify a point.
- Move the cursor away and click at desired location to specify circumferential point of circle. The circle will be created; refer to Figure-39.

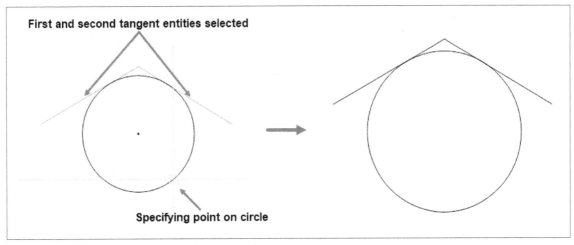

Figure-39. Circle created using tangent entities and point

- Press **Esc** button or click **RMB** to exit the tool.

Creating circle using Tangential, 2 Points tool

Toolbar: Circle > Tangential, 2 Points
Widget: Circle > Tangential, 2 Points
Drop-Down: Circles > Tangential, 2 Points
Menu: Tools > Circle > Tangential, 2 Points

The **Tangential, 2 Points** tool is used to draw a circle tangential to an existing circle and define the diameter and placement by assigning two points on the circumference. The procedure to use this tool is discussed next.

- Click on the **Tangential, 2 Points** tool. You will be asked to select the line, arc, or circle.
- Select desired line, arc, or circle in the Drawing Window as the tangent entity. You will be asked to specify the first point.
- Move the cursor away and click to specify the first point on the circumference of the circle. You will be asked to specify the second point.
- Move the cursor away and click at desired location to specify circumferential point. The circle will be created; refer to Figure-40.
- Press **Esc** button or click **RMB** to exit the tool.

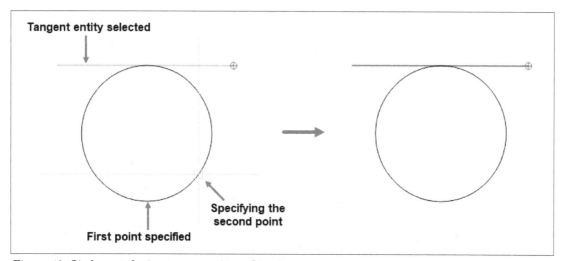

Figure–40. Circle created using tangent entity and 2 points

Creating circle using Tangential 2 Circles, Radius tool

Toolbar: Circle > Tangential 2 Circles, Radius
Widget: Circle > Tangential 2 Circles, Radius
Drop-Down: Circles > Tangential 2 Circles, Radius
Menu: Tools > Circle > Tangential 2 Circles, Radius

The **Tangential 2 Circles, Radius** tool is used to draw a circle tangential to two existing circles with a given radius. The procedure to use this tool is discussed next.

- Click on the **Tangential 2 Circles, Radius** tool. The **Tool Options** will be displayed; refer to Figure-41 and you will be asked to select the first line, arc, or circle.

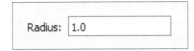

Figure-41. Tangential 2 Circles Radius Tool Options

- Specify desired radius of circle in the **Radius** edit box.
- Select desired line, arc, or circle as a first tangent entity. You will be asked to select the second line, arc, or circle.
- Select desired line, arc, or circle as the second tangent entity. Four centers of the circle will be displayed. You will be asked to select a center point.
- Select desired center for the circle among the four centers defined. The circle will be created; refer to Figure-42.

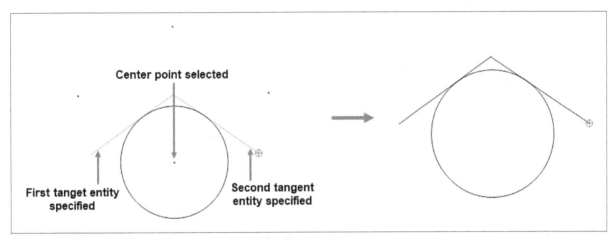

Figure-42. Circle created using 2 tangent entities and radius

- Press **Esc** button or click **RMB** to exit the tool.

Creating circle using Tangential, 3 Circles tool

Toolbar: Circle > Tangential, 3 Circles
Widget: Circle > Tangential, 3 Circles
Drop-Down: Circles > Tangential, 3 Circles
Menu: Tools > Circle > Tangential, 3 Circles

The **Tangential, 3 Circles** tool is used to draw a circle tangential to three existing circles and/or lines. The procedure to use this tool is discussed next.

- Click on the **Tangential, 3 Circles** tool. You will be asked to select the first line, arc, or circle.
- Select desired line, arc, or circle as the first tangent entity. You will be asked to select the second line, arc, or circle.
- Select desired line, arc, or circle as the second tangent entity. You will be asked to select the third line, arc, or circle.
- Select desired line, arc, or circle as the third tangent entity. The six centers are defined to create the circle. You will be asked to select the center.
- Select desired center for the circle among the six centers displayed. The circle will be created; refer to Figure-43.

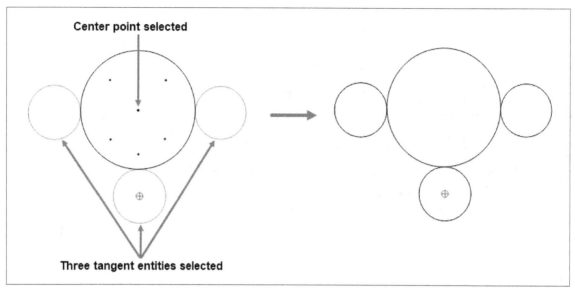

Figure-43. Circle created using 3 tangent entities

- Press **Esc** button or click **RMB** to exit the tool.

CURVE CREATION TOOLS

Curve creation tools are displayed in the **Curve** drop-down; refer to Figure-44. You can also select these tools from **Curve** cascading menu from **Tools** menu; refer to Figure-45 or **Curve** widget; refer to Figure-46 or **Curve** toolbar; refer to Figure-47. These tools are discussed next.

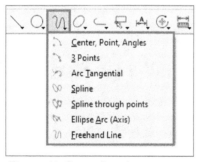

Figure-44. Curve drop down

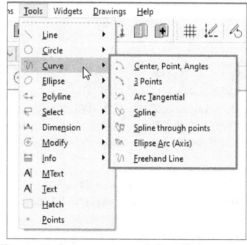

Figure-45. Curve creation tools from Tools menu

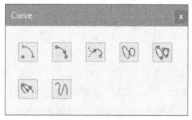

Figure-46. Curve widget

Figure-47. Curve toolbar

Creating curve using Center, Point, Angles tool

Toolbar: Curve > Center, Point, Angles
Widget: Curve > Center, Point, Angles
Drop-Down: Freehand > Center, Point, Angles
Menu: Tools > Curve > Center, Point, Angles

The **Center, Point, Angles** tool is used to draw a curve (arc) with a given radius defined by a center point and a point on the circumference, the direction of rotation (clockwise or counter-clockwise), a point defining the start position of the arc, and a point defining the end position of the arc. The procedure to use this tool is discussed next.

- Click on the **Center, Point, Angles** tool. The **Tool Options** will be displayed; refer to Figure-48 and you will be asked to specify the center of curve.

Figure-48. Center Point Angles Tool Options

- Select desired radio button to set the direction of arc as **Counterclockwise** or **Clockwise**.
- Click in the Drawing Window at desired location to specify the center point of arc. You will be asked to specify the radius of arc.
- Move the cursor away and click at desired location to specify the radius of arc. You will be asked to specify the start angle of arc.
- Move the cursor away and click at desired location to specify the start angle of arc. You will be asked to specify the end angle of the arc.
- Move the cursor away and click at desired location to specify the end angle of the arc. The curve (arc) will be created; refer to Figure-49.

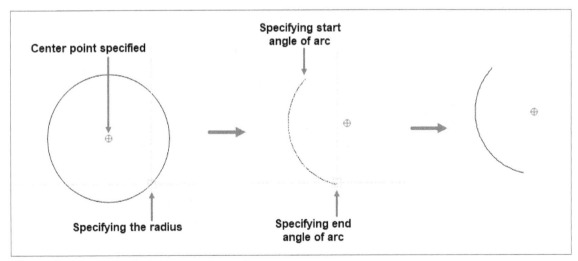

Figure-49. Curve created using center point angles

- Press **Esc** button or click **RMB** to exit the tool.

Creating curve using 3 Points tool

Toolbar: Curve > 3 Points
Widget: Curve > 3 Points
Drop-Down: Freehand > 3 Points
Menu: Tools > Curve > 3 Points

The **3 Points** tool is used to draw a curve (arc) by assigning three points on the circumference of the arc defining the start position, a point on the circumference, and end position of the arc. The procedure to use this tool is discussed next.

- Click on the **3 Points** tool. You will be asked to specify the start point of arc.
- Click in the Drawing Window at desired location to specify start point of arc. You will be asked to specify the second point of arc.
- Move the cursor away and click at desired location to specify the second point of arc. You will be asked to specify the end point of arc.
- Move the cursor away and click at desired location to specify the end point of arc. The arc will be created; refer to Figure-50.

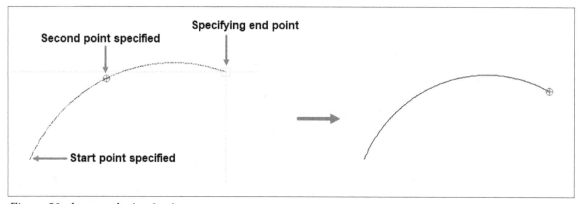

Figure-50. Arc created using 3 points

- Press **Esc** button or click **RMB** to exit the tool.

Creating curve using Arc Tangential tool

Toolbar: Curve > Arc Tangential
Widget: Curve > Arc Tangential
Drop-Down: Freehand > Arc Tangential
Menu: Tools > Curve > Arc Tangential

The **Arc Tangential** tool is used to draw a curve (arc) tangential to the end of an existing line segment with specified radius or angle (deg). The procedure to use this tool is discussed next.

- Click on the **Arc Tangential** tool. The **Tool Options** will be displayed; refer to Figure-51 and you will be asked to select the base entity.

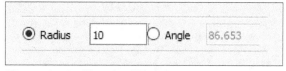

Figure-51. Arc Tangential Tool Options

- The **Radius** radio button is selected by default. Specify desired radius of arc in the **Radius** edit box. Or, select **Angle** radio button and specify desired angle of arc in the **Angle** edit box.
- Select desired base entity from the Drawing Window to create the arc. You will be asked to specify the end angle or end point.
- Specify the end angle if you specified the radius or specify the end point if you specified the angle. The arc will be created; refer to Figure-52.
- Press **Esc** button or click **RMB** to exit the tool.

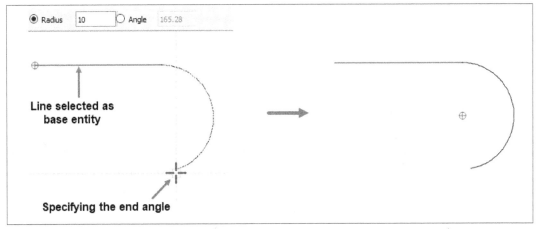

Figure-52. Arc created by defining radius

Creating curve using Spline tool

Toolbar: Curve > Spline
Widget: Curve > Spline
Drop-Down: Freehand > Spline
Menu: Tools > Curve > Spline

The **Spline** tool is used to draw an open or closed spline (curve) by assigning control points and a given degree of freedom (1-3). The procedure to use this tool is discussed next.

- Click on the **Spline** tool. The **Tool Options** will be displayed; refer to Figure-53 and you will be asked to specify the first control point.

Figure-53. Spline Tool Options

- Select desired degree of freedom between **1** and **3** from **Degree** drop-down.
- Select the **Closed** check box to create the closed spline.
- Click on the **Undo** button to undo the process of creating spline.
- Click in the Drawing window at desired location to specify the first control point of spline. You will be asked to specify the second control point.
- Move the cursor away and click at desired location to specify the second control point of spline. You will be asked to specify the third control point.
- Move the cursor away and click at desired location to specify the third control point.
- After specifying the third control point, click **RMB**. The spline will be created; refer to Figure-54. Note that, in case of specifying the degree of freedom as **3**, you will be asked to specify upto four control points to create the spline.

- Press **Esc** button or click **RMB** to exit the tool.

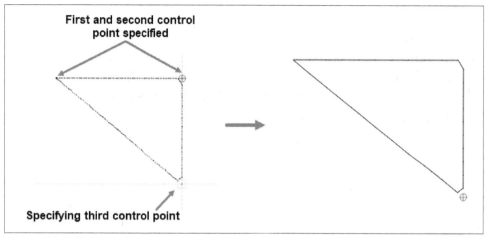

Figure-54. Spline created

Creating curve using Spline through points tool

Toolbar: Curve > Spline through points
Widget: Curve > Spline through points
Drop-Down: Freehand > Spline through points
Menu: Tools > Curve > Spline through points

The **Spline through points** tool is used to draw an open or closed spline (curve) by defining points on the spline. The procedure to use this tool is discussed next.

- Click on the **Spline through points** tool. The **Tool Options** will be displayed; refer to Figure-55 and you will be asked to specify the first control point.

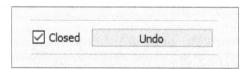

Figure-55. Spline through points Tool Options

- Select **Closed** check box to create the closed spline.
- Click on the **Undo** button to undo the process of creating spline.
- Click in the Drawing Window at desired location to specify the first control point. You will be asked to specify the next control point(s).
- Move the cursor away and click at desired location to specify the next control point(s).
- After specifying the point(s), click **RMB**. The spline will be created; refer to Figure-56.

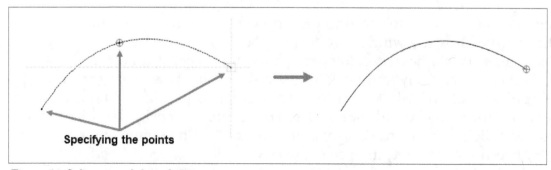

Figure-56. Spline created through points

• Press **Esc** button or click **RMB** to exit the tool.

Creating curve using Ellipse Arc (Axis) tool

Toolbar: Curve > Ellipse Arc (Axis)
Widget: Curve > Ellipse Arc (Axis)
Drop-Down: Freehand > Ellipse Arc (Axis)
Menu: Tools > Curve > Ellipse Arc (Axis)

The **Ellipse Arc** tool is used to draw an elliptical arc by defining points of ellipse. The procedure to use this tool is discussed next.

• Click on the **Ellipse Arc (Axis)** tool. You will be asked to specify the ellipse center.
• Click in the Drawing Window at desired location to specify the center of ellipse. You will be asked to specify endpoint of major axis.
• Move the cursor away and click at desired location to specify the endpoint of major axis. You will be asked to specify the endpoint or length of minor axis.
• Move the cursor away and click at desired location to specify the endpoint or length of minor axis. You will be asked to specify the start angle.
• Move the cursor away and click at desired location to specify the start angle of arc. You will be asked to specify the end angle.
• Move the cursor away and click at desired location to specify the end angle of arc. The ellipse arc will be created; refer to Figure-57.

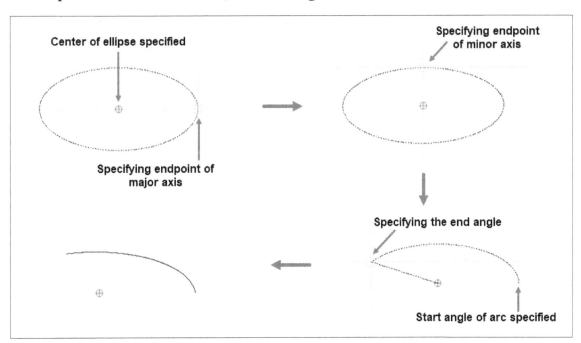

Figure-57. Ellipse arc created

• Press **Esc** button or click **RMB** to exit the tool.

Creating curve using Freehand Line tool

Toolbar: Curve > Freehand Line
Widget: Curve > Freehand Line
Drop-Down: Curve > Freehand Line
Menu: Tools > Curve > Freehand Line

The **Freehand Line** tool is used to draw a non-geometric line. The procedure to use this tool is discussed next.

- Click on the **Freehand** tool. You will be asked to draw a line.
- Press and hold the **LMB** and Move the cursor at desired location to create the freehand line. The freehand line will be created; refer to Figure-58.

Figure-58. Freehand line created

- Press **Esc** button or click **RMB** to exit the tool.

ELLIPSE CREATION TOOLS

Ellipse creation tools are displayed in the **Ellipses** drop-down; refer to Figure-59. You can also select these tools from **Ellipse** cascading menu of **Tools** menu; refer to Figure-60 or **Ellipse** widget; refer to Figure-61 or **Ellipse** toolbar; refer to Figure-62. These tools are discussed next.

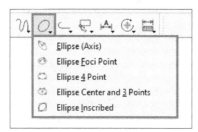

Figure-59. Ellipses drop down

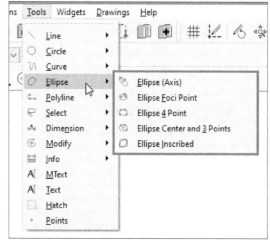

Figure-60. Ellipse creation tools from Tools menu

Figure-61. Ellipse widget

Figure-62. Ellipse toolbar

Ellipse (Axis)

Toolbar: Ellipse > Ellipse (Axis)
Widget: Ellipse > Ellipse (Axis)
Drop-Down: Ellipses > Ellipse (Axis)
Menu: Tools > Ellipse > Ellipse (Axis)

The **Ellipse (Axis)** tool is used to draw an ellipse by assigning a center point, a point on the circumference of major axis, and a point on the circumference of the minor axis. The procedure to use this tool is discussed next.

- Click on the **Ellipse (Axis)** tool. You will be asked to specify the center of ellipse.
- Click in the Drawing Window at desired location to specify the center point of ellipse. You will be asked to specify the endpoint of major axis.
- Move the cursor away and click at desired location and specify the endpoint of major axis. You will be asked to specify the endpoint or length of minor axis.
- Move the cursor away and click at desired location to specify the endpoint or length of the minor axis. The ellipse will be created; refer to Figure-63.

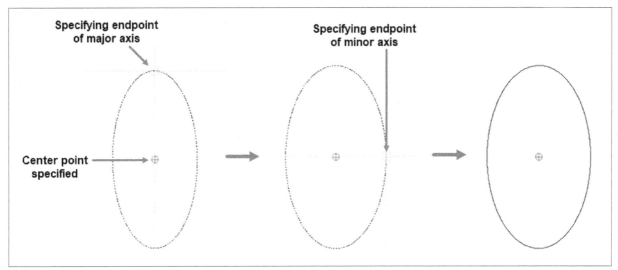

Figure-63. Ellipse created using axes

- Press **Esc** button or click **RMB** to exit the tool.

Ellipse Foci Point

Toolbar: Ellipse > Ellipse Foci Point
Widget: Ellipse > Ellipse Foci Point
Drop-Down: Ellipses > Ellipse Foci Point
Menu: Tools > Ellipse > Ellipse Foci Point

The **Ellipse Foci Point** tool is used to draw an ellipse by assigning two foci points and a point on the circumference. The procedure to use this tool is discussed next.

- Click on the **Ellipse Foci Point** tool. You will be asked to specify the first focus of ellipse.
- Click in the Drawing Window at desired location to specify the first focus of ellipse. You will be asked to specify the second focus of ellipse.

- Move the cursor away and click at desired location to specify the second focus of ellipse. You will be asked to specify a point on ellipse or total distance to foci.
- Move the cursor away and click at desired location to specify a point on ellipse or total distance to foci. The ellipse will be created; refer to Figure-64.

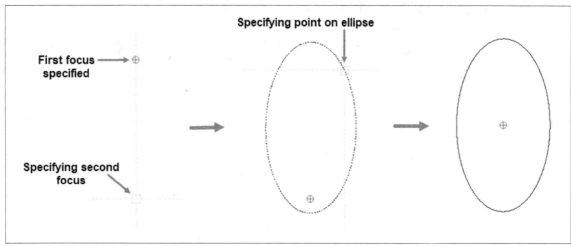

Figure-64. Ellipse created using foci points

- Press **Esc** button or click **RMB** to exit the tool.

Ellipse 4 Point

Toolbar: Ellipse > Ellipse 4 Point	
Widget: Ellipse > Ellipse 4 Point	
Drop-Down: Ellipses > Ellipse 4 Point	
Menu: Tools > Ellipse > Ellipse 4 Point	

The **Ellipse 4 Point** tool is used to draw an ellipse assigning four points on the circumference. The procedure to use this tool is discussed next.

- Click on the **Ellipse 4 Point** tool. You will be asked to specify the first point on ellipse.
- Click in the Drawing Window at desired location to specify the first point on ellipse. You will be asked to specify the second point on ellipse.
- Move the cursor away and click at desired location to specify the second point on ellipse. You will be asked to specify the third point on ellipse.
- Move the cursor away and click at desired location to specify the third point on ellipse. You will be asked to specify the fourth point on ellipse.
- Move the cursor away and click at desired location to specify the fourth point on ellipse. The ellipse will be created; refer to Figure-65.

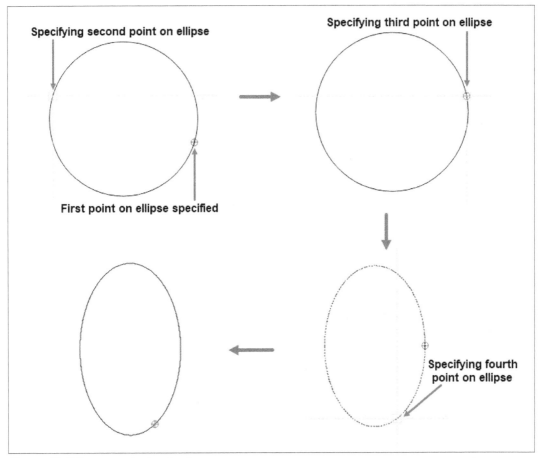

Figure-65. Ellipse created using 4 points

- Press **Esc** button or click **RMB** to exit the tool.

Ellipse Center and 3 Points

Toolbar: Ellipse > Ellipse Center and 3 Points
Widget: Ellipse > Ellipse Center and 3 Points
Drop-Down: Ellipses > Ellipse Center and 3 Points
Menu: Tools > Ellipse > Ellipse Center and 3 Points

The **Ellipse Center and 3 Points** tool is used to draw an ellipse by assigning a center point of ellipse and three points on the circumference of ellipse. The procedure to use this tool is discussed next.

- Click on the **Ellipse Center and 3 Points** tool. You will be asked to specify the center of ellipse.
- Click in the Drawing Window at desired location to specify the center of ellipse. You will be asked to specify the first point on ellipse.
- Move the cursor away and click at desired location to specify the first point on ellipse. You will be asked to specify the second point on ellipse.
- Move the cursor away and click at desired location to specify the second point on ellipse. You will be asked to specify the third point on ellipse.
- Move the cursor away and click at desired location to specify the third point on ellipse. The ellipse will be created; refer to Figure-66.

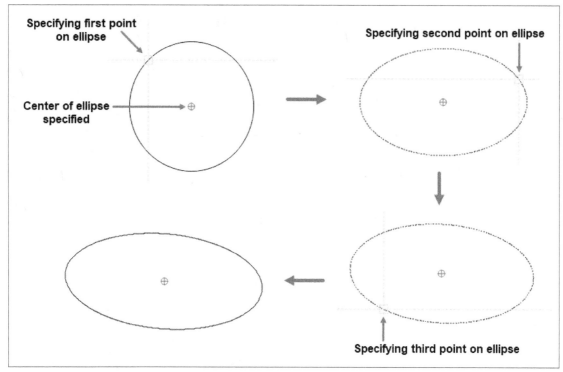

Figure-66. Ellipse created using center and 3 points

- Press **Esc** button or click **RMB** to exit the tool.

Ellipse Inscribed

Toolbar: Ellipse > Ellipse Inscribed
Widget: Ellipse > Ellipse Inscribed
Drop-Down: Ellipses > Ellipse Inscribed
Menu: Tools > Ellipse > Ellipse Inscribed

The **Ellipse Inscribed** tool is used to draw an ellipse constrained by four existing non-parallel line segments. The procedure to use this tool is discussed next.

- Click on the **Ellipse Inscribed** tool. You will be asked to select the first line.
- Select desired line in the Drawing Window as first line to create ellipse. You will be asked to select the second line.
- Select desired line as second line to create ellipse. You will be asked to select the third line.
- Select desired line as third line to create ellipse. You will be asked to select the fourth line.
- As you hover the cursor on fourth line, the preview of ellipse will be displayed.
- Click on the fourth line. The ellipse will be created; refer to Figure-67.

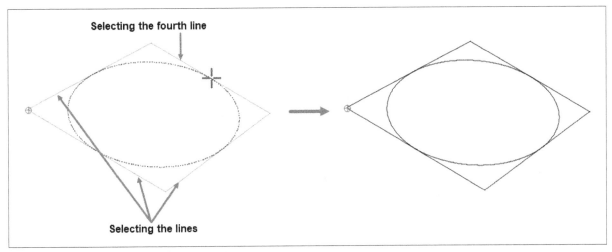

Figure-67. Ellipse created using lines

- Press **Esc** button or click **RMB** to exit the tool.

POLYLINE CREATION TOOLS

Polyline creation tools are displayed in the **Polylines** drop-down; refer to Figure-68. You can also select these tools from **Polyline** cascading menu from **Tools** menu; refer to Figure-69 or **Polyline** widget; refer to Figure-70 or **Polyline** toolbar; refer to Figure-71. These tools are discussed next.

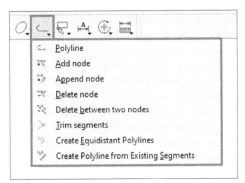

Figure-68. Polylines drop down

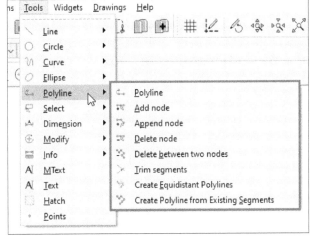

Figure-69. Polyline creation tools from Tools menu

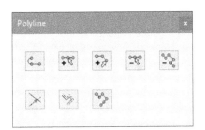

Figure-70. Polyline widget

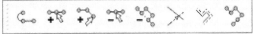

Figure-71. Polyline toolbar

Polyline

Toolbar: Polyline > Polyline
Widget: Polyline > Polyline
Drop-Down: PolyLines > Polyline
Menu: Tools > Polyline > Polyline

The **Polyline** tool is used to draw an open or closed continuous line consisting of one or more straight line or arc segments defined by endpoints and/or radius or angle for arcs. The procedure to use this tool is discussed next.

- Click on the **Polyline** tool. The **Tool Options** will be displayed; refer to Figure-72 and you will be asked to specify the first point.

Figure-72. Polyline Tool Options

- Click on the **Close** button to close the polyline.
- Click on the **Undo** button to undo the process of creating polyline.
- Select desired option from drop-down in the **Tool Options**. Select the **Line** option from the drop-down to create the polyline using points. Select the **Tangential** option from the drop-down to create the line and arc using points. Select the **Tan Radius** option from the drop-down create the line and arc using points and radius. Select the **Angle** option from the drop-down to create clockwise or counter clockwise arc using points and angle.
- Click in the Drawing Window at desired location to specify the first point. You will be asked to specify the next point(s).
- Move the cursor away and click at desired location to specify the next point(s).
- After specifying desired number of point(s), click **RMB** to create the polyline. The polyline will be created; refer to Figure-73.

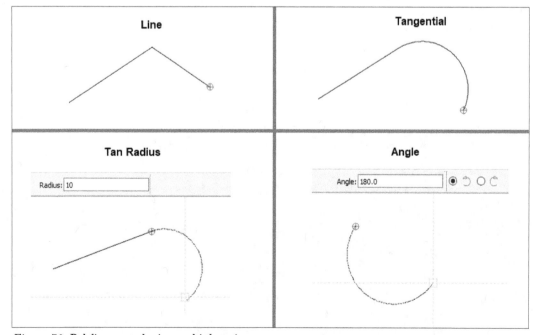

Figure-73. Polyline created using multiple options

- Press **Esc** button or click **RMB** again to exit the tool.

Add node

Toolbar: Polyline > Add node
Widget: Polyline > Add node
Drop-Down: PolyLines > Add node
Menu: Tools > Polyline > Add node

The **Add node** tool is used to add node to existing polyline. The procedure to use this tool is discussed next.

- Click on the **Add node** tool. You will be asked to select the polyline to add nodes.
- Select **Snap on Entity** button from **Snap Selection** toolbar. The object snapping will be activated.
- Select desired polyline to which you want to add node(s) and click on the polyline at desired location(s) to add the node point(s). The node point(s) will be added; refer to Figure-74.
- Press **Esc** button or click **RMB** to exit the tool.

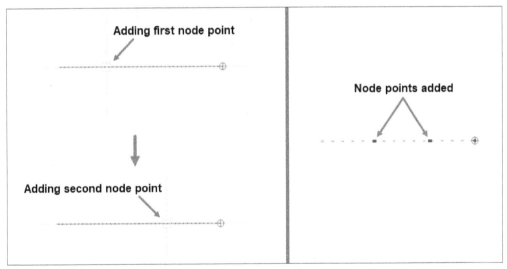

Figure-74. Node points added

- Deselect the **Snap on Entity** button from **Snap Selection** toolbar. Now, the polyline can be stretched at desired location(s) using node point(s); refer to Figure-75.

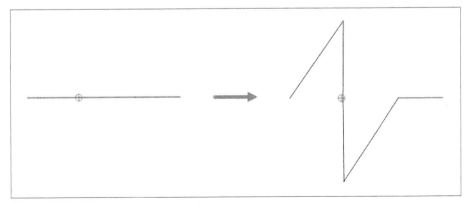

Figure-75. Polyline stretched using node points

Append node

Toolbar: Polyline > Append node
Widget: Polyline > Append node
Drop-Down: PolyLines > Append node
Menu: Tools > Polyline > Append node

The **Append node** tool is used to add one or more segments to an existing polyline by selecting polyline and adding new node endpoint. The procedure to use this tool is discussed next.

- Click on the **Append node** tool. You will be asked to select the polyline.
- Select desired polyline at or near the start or end point of the polyline. A node point will be added to the start or end point of that polyline, respectively and a segment will be attached to that node point. You will be asked to specify the next point(s) of segment.
- Move the cursor away and click at desired location(s) to specify the next point(s) of segment.
- After specifying desired number of point(s), click **RMB**. The segment(s) will be added to the polyline; refer to Figure-76.

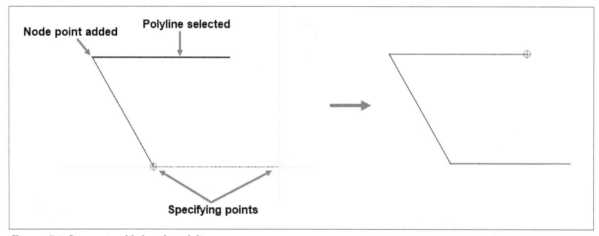

Figure-76. Segments added to the polyline

- Press **Esc** button or click **RMB** to exit the tool.

Delete node

Toolbar: Polyline > Delete node
Widget: Polyline > Delete node
Drop-Down: PolyLines > Delete node
Menu: Tools > Polyline > Delete node

The **Delete node** tool is used to delete selected node of an existing polyline. The procedure to use this tool is discussed next.

- Click on the **Delete node** tool. You will be asked to select the polyline to delete node.
- Select desired polyline from which you want to delete node(s). You will be asked to specify the node point(s).

- Click on the node point(s) on the polyline which you want to delete. The node point(s) will be deleted; refer to Figure-77.

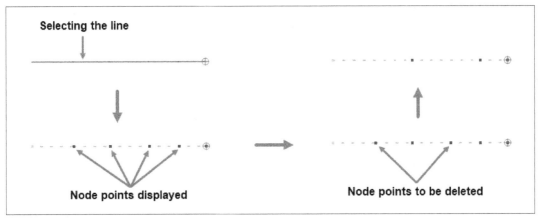

Figure-77. Node points deleted

- Press **Esc** button or click **RMB** to exit the tool.

Delete between two nodes

Toolbar: Polyline > Delete between two nodes
Widget: Polyline > Delete between two nodes
Drop-Down: PolyLines > Delete between two nodes
Menu: Tools > Polyline > Delete between two nodes

The **Delete between two nodes** tool is used to delete one or more nodes between selected nodes of an existing polyline. The procedure to use this tool is discussed next.

- Select desired polyline from which you want to delete the node(s) between the selected nodes and click on the **Delete between two nodes** tool. You will be asked to select the polyline.
- Select that polyline again. You will be asked to select the first node and second node.
- Select the first node and then select the second node between which you want to delete the node(s). The node(s) will be deleted; refer to Figure-78.

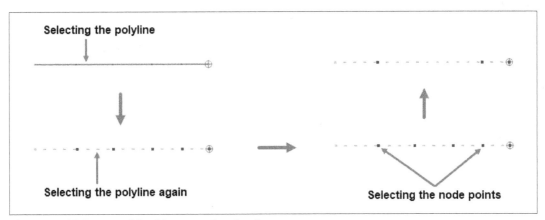

Figure-78. Node points deleted between selected nodes

- Press **Esc** button or click **RMB** to exit the tool.

Trim segments

Toolbar: Polyline > Trim segments
Widget: Polyline > Trim segments
Drop-Down: PolyLines > Trim segments
Menu: Tools > Polyline > Trim segments

The **Trim segments** tool is used to extend two separate non-parallel segments of an existing polyline to intersect at a new node or trim the extra portion of polyline. The procedure to use this tool is discussed next.

- Click on the **Trim segments** tool. You will be asked to select the polyline.
- Select desired polyline to intersect the segments. You will be asked to select the first and second segment to extend.
- Select the first and second segment which you want to extend. The segment will be extended; refer to Figure-79.

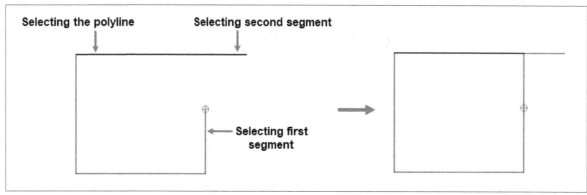

Figure-79. Segment extended

- Press **Esc** button or click **RMB** to exit the tool.

Create Equidistant Polylines

Toolbar: Polyline > Create Equidistant Polylines
Widget: Polyline > Create Equidistant Polylines
Drop-Down: PolyLines > Create Equidistant Polylines
Menu: Tools > Polyline > Create Equidistant Polylines

The **Create Equidistant Polylines** tool is used to draw a given number of polylines parallel to a selected existing polyline with a given distance between lines. The procedure to use this tool is discussed next.

- Click on the **Create Equidistant Polylines** tool. The **Tool Options** will be displayed; refer to Figure-80 and you will be asked to select the original polyline.

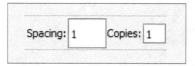

Figure-80. Create Equidistant Polylines Tool Options

- Specify desired distance of polyline to be created from the original polyline in the **Spacing** edit box.

- Specify desired number of copies to be created in the **Copies** edit box.
- After specifying desired parameters, select the original polyline to create equidistant parallel polylines. The parallel polylines will be created; refer to Figure-81.
- Press **Esc** button or click **RMB** to exit the tool.

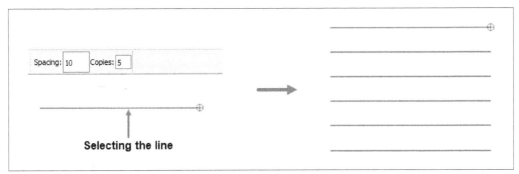

Figure-81. Parallel polylines created

Create Polyline from Existing Segments

Toolbar: Polyline > Create Polyline from Existing Segments
Widget: Polyline > Create Polyline from Existing Segments
Drop-Down: PolyLines > Create Polyline from Existing Segments
Menu: Tools > Polyline > Create Polyline from Existing Segments

The **Create Polyline from Existing Segments** tool is used to create polyline from two or more existing separate line or arc segments forming a continuous line. The procedure to use this tool is discussed next.

- Click on the **Create Polyline from Existing Segments** tool. You will be asked to select two or more separate segments.
- Click on the two or more separate segments to create continuous polyline. The separate segments will be converted into a continuous polyline; refer to Figure-82.

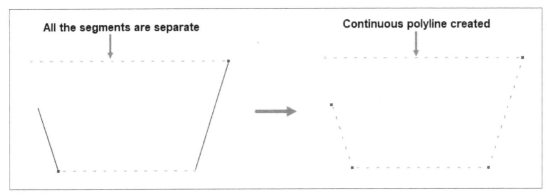

Figure-82. Continuous polyline created

- Press **Esc** button or click **RMB** to exit the tool.

SELECTION TOOLS

Selection tools are displayed in the **Select** drop-down; refer to Figure-83. You can also select these tools from **Select** cascading menu from **Tools** menu; refer to Figure-84 or **Select** widget; refer to Figure-85 or **Select** toolbar; refer to Figure-86. These tools are discussed next.

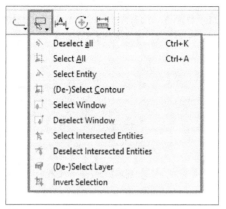

Figure-83. Select drop down

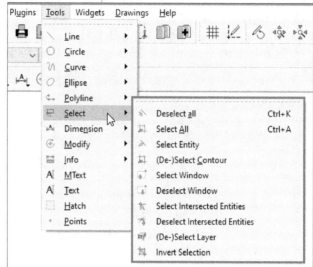

Figure-84. Select creation tools from Tools menu

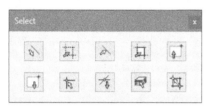

Figure-85. Select widget

Figure-86. Select toolbar

Deselect all

Toolbar:	Select > Deselect all
Widget:	Select > Deselect all
Drop-Down:	Select > Deselect all
Menu:	Tools > Select > Deselect all

The **Deselect all** tool is used to deselect all entities on visible layers; refer to Figure-87.

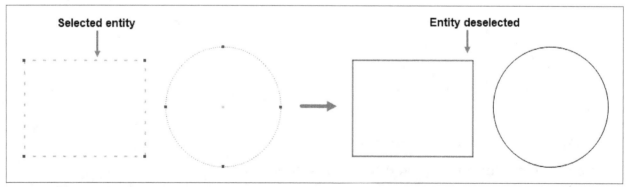

Figure-87. Entity deselected

Select All

Toolbar: Select > Select All
Widget: Select > Select All
Drop-Down: Select > Select All
Menu: Tools > Select > Select All

The **Select All** tool is used to select all entities on visible layers; refer to Figure-88.

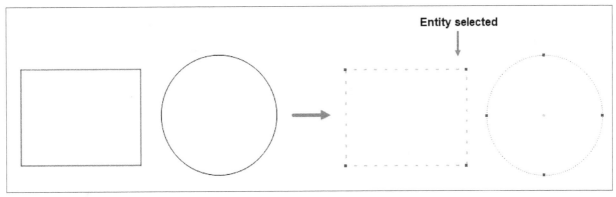

Figure-88. Entity selected

Select Entity

Toolbar: Select > Select Entity
Widget: Select > Select Entity
Drop-Down: Select > Select Entity
Menu: Tools > Select > Select Entity

The **Select Entity** tool is used to select, or deselect, one or more entities.

(De-)Select Contour

Toolbar: Select > (De-)Select Contour
Widget: Select > (De-)Select Contour
Drop-Down: Select > (De-)Select Contour
Menu: Tools > Select > (De-)Select Contour

The **(De-)Select Contour** tool is used to select or deselect entities connected by shared points; refer to Figure-89.

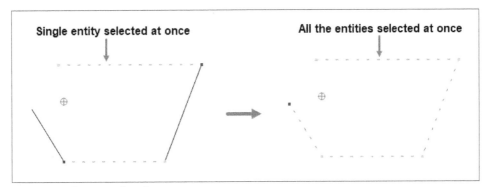

Figure-89. Contour selected

Select Window

Toolbar: Select > Select Window
Widget: Select > Select Window
Drop-Down: Select > Select Window
Menu: Tools > Select > Select Window

The **Select Window** tool is used to select one or more entities enclosed by selection window (Left to Right), or crossed by selection window (Right to Left); refer to Figure-90.

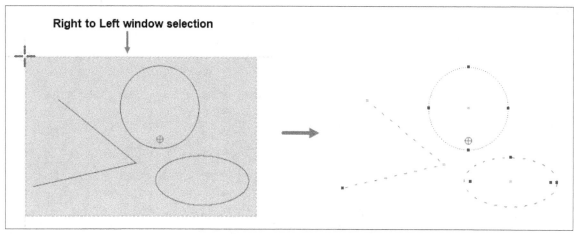

Figure-90. Entities selected by window selection

Deselect Window

Toolbar: Select > Deselect Window
Widget: Select > Deselect Window
Drop-Down: Select > Deselect Window
Menu: Tools > Select > Deselect Window

The **Deselect Window** tool is used to deselect one or more entities enclosed by selection window (Left to Right), or crossed by selection window (Right to Left); refer to Figure-91.

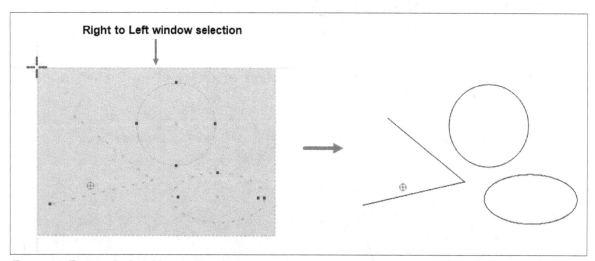

Figure-91. Entities deselected by window selection

Select Intersected Entities

Toolbar: Select > Select Intersected Entities
Widget: Select > Select Intersected Entities
Drop-Down: Select > Select Intersected Entities
Menu: Tools > Select > Select Intersected Entities

The **Select Intersected Entities** tool is used to select one or more entities crossed by selection line; refer to Figure-92.

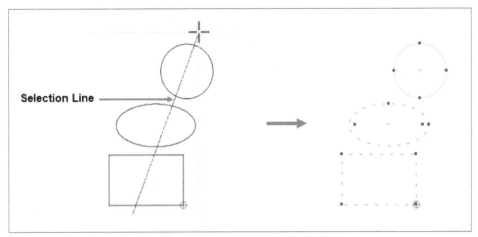

Figure-92. Entities selected by selection line

Deselect Intersected Entities

Toolbar: Select > Deselect Intersected Entities
Widget: Select > Deselect Intersected Entities
Drop-Down: Select > Deselect Intersected Entities
Menu: Tools > Select > Deselect Intersected Entities

The **Deselect Intersected Entities** tool is used to deselect one or more entities crossed by selection line; refer to Figure-93.

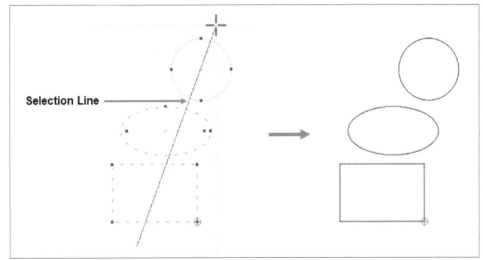

Figure-93. Entities deselected by selection line

(De-)Select Layer

Toolbar: Select > (De-)Select Layer
Widget: Select > (De-)Select Layer
Drop-Down: Select > (De-)Select Layer
Menu: Tools > Select > (De-)Select Layer

The **(De-)Select Layer** tool is used to select or deselect all entities on the layer of the selected entity; refer to Figure-94.

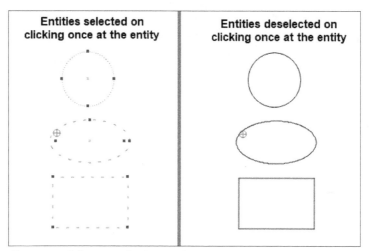

Figure-94. Selecting or deselecting the entities at once

Invert Selection

Toolbar: Select > Invert Selection
Widget: Select > Invert Selection
Drop-Down: Select > Invert Selection
Menu: Tools > Select > Invert Selection

The **Invert Selection** tool is used to select all un-selected entities or de-select all selected entities.

SELF ASSESSMENT

Q1. Which of the following tools is used to draw a line tangent to an existing circle and perpendicular to an existing line?

a) Tangent Orthogonal
b) Tangent (C,C)
c) Tangent (P,C)
d) Orthogonal

Q2. Which of the following tools is used to draw a polygon with a given number of sides assigning the center point and point of center of one side?

a) Polygon (Cen,Cor)
b) Polygon (Cen,Tan)
c) Polygon (Cor,Cor)
d) Polygon (Cen,Cen)

Q3. Which of the following tools is used to draw an ellipse by assigning a center point, a point on the circumference of major axis, and a point on the circumference of the minor axis?

a) Ellipse (Axis)
b) Ellipse Foci Point
c) Ellipse 4 Point
d) Ellipse Center and 3 Points

Q4. The **Angle** tool is used to draw a line from an assigned point defining the start, middle, or end of the line and with an assigned   and   .

Q5. The   tool is used to draw a line with a given length and at a given angle relative to an existing line placing the center of the line at an assigned point.

Q6. The   tool is used to draw a curve (arc) with specified radius defined by a center point and a point on the circumference, the direction of rotation (clockwise or counter-clockwise), a point defining the start position of the arc, and a point defining the end position of the arc.

Q7. The **Append node** tool is used to add one or more segments to an existing polyline by selecting polyline and adding new node endpoint. (True/False)

Q8. The **Polyline** tool is used to draw a given number of polylines parallel to a selected existing polyline with a given distance between lines. (True/False)

Q9. The **(De-)Select Contour** tool is used to select or deselect entities connected by shared points. (True/False)

FOR STUDENT NOTES

Ans. **1.** Tangent Orthogonal **2.** Polygon (Cen,Tan) **3.** Ellipse (Axis) **4.** length and angle **5.** Relative Angle **6.** Center, Point, Angles **7.** True **8.** False **9.** True

Chapter 3

Advanced Dimensioning and Practice

Topics Covered

The major topics covered in this chapter are:

- *Dimensioning and its relations with drawing*
- *Dimension Style*
- *Practical 1*
- *Practical 2*
- *Practice Drawings*

DIMENSIONING AND ITS RELATIONS

When we create a dimension in a sketch it is not confined to that sketch only. You will learn later that it also affects the parameters in 2D model and drafting environment. At that time, the style and dimensions that we apply in sketch will be reflected in the drawing. So, it is very important to understand dimension styles here. In the previous chapters, we have worked on basic dimensions. In this chapter, we will discuss the dimensions and styles in detail.

DIMENSION TOOLS

Dimension tools are displayed in the **Dimensions** drop-down; refer to Figure-1. You can also select these tools from **Dimension** cascading menu of **Tools** menu; refer to Figure-2 or **Dimension** widget; refer to Figure-3 or **Dimension** toolbar; refer to Figure-4. These tools are discussed next.

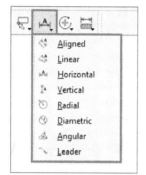

Figure-1. Dimensions drop down

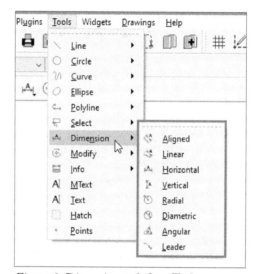

Figure-2. Dimension tools from Tools menu

Figure-3. Dimension widget

Figure-4. Dimension toolbar

Aligned

Toolbar: Dimension > Aligned
Widget: Dimension > Aligned
Drop-Down: Dimension > Aligned
Menu: Tools > Dimension > Aligned

The **Aligned** tool is used to align dimension lines and text aligned to an existing entity. The procedure to use this tool is discussed next.

• Click on the **Aligned** tool. The **Tool Options** will be displayed; refer to Figure-5. You will be asked to select origin point of line.

Figure-5. Aligned Tool Options

- Click on the  button to toggle between the diameter symbol.
- Click on the ⊘ ∨ drop-down and select desired symbol from the list.
- Specify desired text in the edit boxes of **Tool Options**.
- After specifying desired parameters, click at the origin point of aligned line. The dimension line will be attached to cursor and you will be asked to select the end point of line.
- Move the cursor away and click at the end point of the aligned line. The aligned dimension will be attached to cursor.
- Move the cursor away and click at desired location to place the dimension. The aligned dimension will be applied; refer to Figure-6.

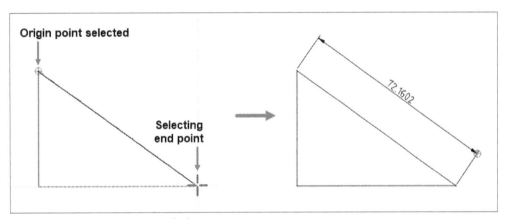

Figure-6. Aligned dimension applied

- Press **Esc** button or click **RMB** to exit the tool.

Linear

Toolbar: Dimension > Linear	
Widget: Dimension > Linear	
Drop-Down: Dimension > Linear	
Menu: Tools > Dimension > Linear	

The **Linear** tool is used to apply dimension lines and text at the defined angle to an entity by selecting start and end points on a line segment and placement point for the text. The procedure to use this tool is discussed next.

- Click on the **Linear** tool. The **Tool Options** will be displayed; refer to Figure-7 and you will be asked to select origin point of line.

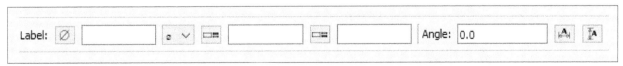

Figure-7. Linear Tool Options

- Specify desired angle of dimension line in the **Angle** edit box.
- Click on the button to apply the dimension on horizontal entity.

- Click on the 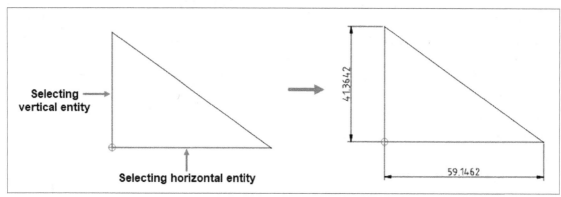 button to apply the dimension on vertical entity.
- After specifying desired parameters, click at the origin point of linear line. The dimension line will be attached to cursor and you will be asked to select the end point of line.
- Move the cursor away and click at the end point of the linear line. The linear dimension will be attached to cursor.
- Move the cursor away and click at desired location to place the dimension. The linear dimension will be applied; refer to Figure-8.

Figure-8. Linear dimensions applied

- Press **Esc** button or click **RMB** to exit the tool.

Horizontal

Toolbar: Dimension > Horizontal
Widget: Dimension > Horizontal
Drop-Down: Dimension > Horizontal
Menu: Tools > Dimension > Horizontal

The **Horizontal** tool is used to apply dimension lines and text aligned to an entity by selecting start and end points on a line segment and placement point for the text. The procedure to use this tool is same as discussed earlier.

- Click on the **Horizontal** tool. The **Tool Options** will be displayed; refer to Figure-9.

Figure-9. Horizontal Tool Options

- The parameters in the **Tool Options** has been discussed earlier.
- After specifying desired parameters, follow the procedure as discussed earlier. The horizontal dimension will be applied; refer to Figure-10.

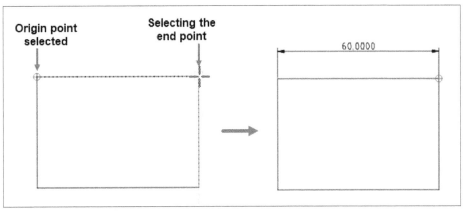

Figure-10. Horizontal dimension applied

- Press **Esc** button or click **RMB** to exit the tool.

Vertical

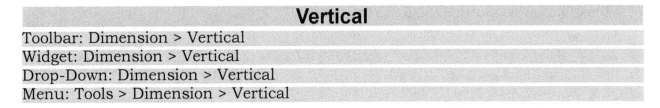

Toolbar: Dimension > Vertical
Widget: Dimension > Vertical
Drop-Down: Dimension > Vertical
Menu: Tools > Dimension > Vertical

The **Vertical** tool is used to apply dimension lines and text aligned to an entity by selecting start and end points on a line segment and placement point for the text. The procedure to use this tool is same as discussed earlier; refer to Figure-11.

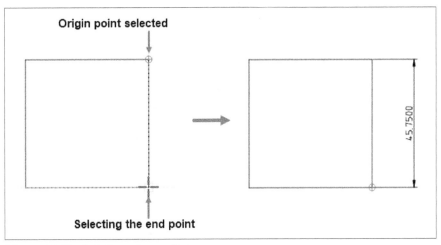

Figure-11. Vertical dimension applied

Radial

Toolbar: Dimension > Radial
Widget: Dimension > Radial
Drop-Down: Dimension > Radial
Menu: Tools > Dimension > Radial

The **Radial** dimension tool is used to apply radius dimension to circles and arcs. The procedure to use this tool is discussed next.

- Click on the **Radial** tool. The **Tool Options** will be displayed as discussed earlier and you will be asked to select the arc or circle entity.

- Select desired arc or circle to which you want to apply the radial dimension. The radial dimension will be applied; refer to Figure-12.

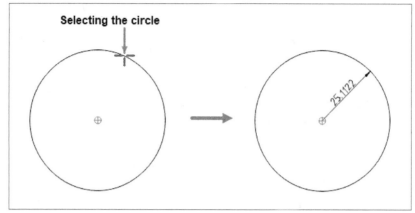

Figure-12. Radial dimension applied

- Press **Esc** button or click **RMB** to exit the tool.

Diametric

Toolbar: Dimension > Diametric

Widget: Dimension > Diametric

Drop-Down: Dimension > Diametric

Menu: Tools > Dimension > Diametric

The **Diametric** tool is used to diameter dimension to arc/circle. The procedure to use this tool is same as discussed for **Radial** tool; refer to Figure-13.

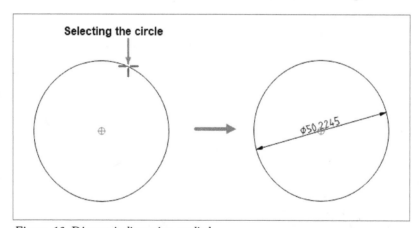

Figure-13. Diametric dimension applied

Angular

Toolbar: Dimension > Angular

Widget: Dimension > Angular

Drop-Down: Dimension > Angular

Menu: Tools > Dimension > Angular

The **Angular** tool is used to apply angular dimension by selecting two existing non-parallel line segments and placement point for the text. The procedure to use this tool is discussed next.

- Click on the **Angular** tool. The **Tool Options** will be displayed as discussed earlier and you will be asked to select first line.
- Select desired line in the Drawing Window. You will be asked to select the second line.
- Select the second adjacent line. The angular dimension will be attached to the cursor.
- Move the cursor away and click at desired location to place the dimension. The angular dimension will be applied; refer to Figure-14.

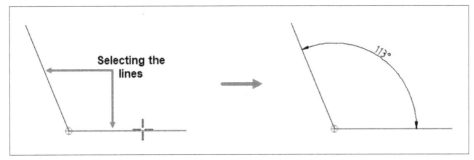

Figure-14. Angular dimension applied

- Press **Esc** button or click **RMB** to exit the tool.

Leader

Toolbar: Dimension > Leader
Widget: Dimension > Leader
Drop-Down: Dimension > Leader
Menu: Tools > Dimension > Leader

The **Leader** tool is used to draw a text leader by selecting start (arrow location), intermediate, and end points. The procedure to use this tool is discussed next.

- Click on the **Leader** tool. You will be asked to select a target point.
- Click in the Drawing Window at desired location to specify the target point. You will be asked to specify the next point(s).
- Move the cursor away and click at desired location to specify the next point(s).
- After specifying desired number of points, press **Enter** button or click **RMB**. The leader will be created; refer to Figure-15.

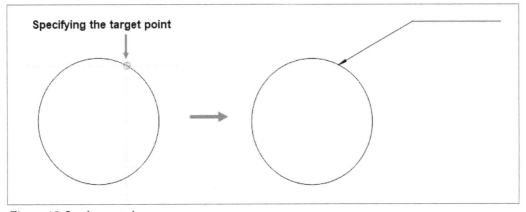

Figure-15. Leader created

- Press **Esc** button or click **RMB** to exit the tool.

MODIFICATION TOOLS

Modification tools are displayed in the **Modify** drop-down; refer to Figure-16. You can also select these tools from **Modify** cascading menu of **Tools** menu; refer to Figure-17 or **Modify** widget; refer to Figure-18 or **Modify** toolbar; refer to Figure-19. These tools are discussed next.

Figure-16. Modify drop down

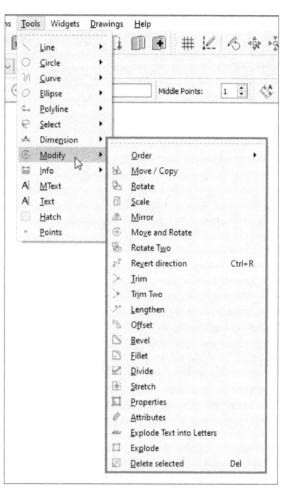

Figure-17. Modification tools from Tools menu

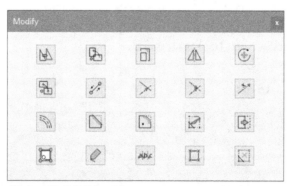

Figure-18. Modify widget

Figure-19. Modify toolbar

Move/Copy

Toolbar: Modify > Move/Copy
Widget: Modify > Move/Copy
Drop-Down: Modify > Move/Copy
Menu: Tools > Modify > Move/Copy

The **Move/Copy** tool is used to move selected entity by defining a reference point and a relative target point. Optionally, keep the original entity (Copy), create multiple copies and/or alter attributes and layer. The procedure to use this tool is discussed next.

- Click on the **Move/Copy** tool. You will be asked to select the entity to move.
- Select desired entity which you want to move and press **Enter**. You will be asked to specify the reference point.
- Click in the Drawing Window at desired location to specify the reference point. You will be asked to specify the target point.
- Move the cursor away and click at desired location to specify the target point of entity. The **Move/Copy Options** dialog box will be displayed; refer to Figure-20.

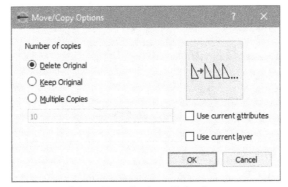

Figure-20. Move Copy Options dialog box

- Select desired radio button from **Number of copies** area of the dialog box. Select **Delete Original** radio button to delete the original entity after moving or copying the entity. Select **Keep Original** radio button to keep the original entity after moving or copying the entity. Select **Multiple Copies** radio button to create multiple copies of an entity by specifying the value in the edit box.
- Select **Use current attributes** check box to use the current specified parameters.
- Select **Use current layer** check box to use the current layer.
- After specifying desired parameters, click on the **OK** button from the dialog box. The entity will be moved or copied; refer to Figure-21.
- Press **Esc** button or click **RMB** to exit the tool.

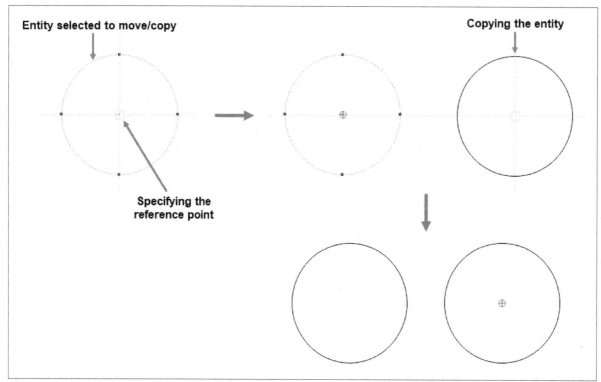

Figure-21. Entity copied

Rotate

Toolbar: Modify > Rotate
Widget: Modify > Rotate
Drop-Down: Modify > Rotate
Menu: Tools > Modify > Rotate

The **Rotate** tool is used to rotate selected entity about a rotation point, moving the entity from a reference point to a target point. The procedure to use this tool is discussed next.

- Click on the **Rotate** tool. You will be asked to select the entity to rotate.
- Select desired entity in the Drawing Window which you want to rotate and press **Enter** button. You will be asked to specify the center of rotation.
- Click in the Drawing Window to specify the center of rotation. You will be asked to specify the reference point.
- Click in the Drawing Window at desired location to specify the reference point for rotation. You will be asked to specify the target point to rotate.
- Move the cursor away and click at desired location to specify the target point of rotation. The **Rotation Options** dialog box will be displayed; refer to Figure-22.

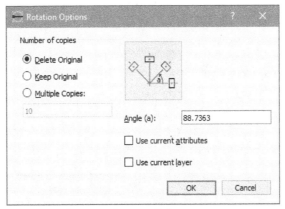

Figure-22. Rotation Options dialog

- Specify desired angle of rotation in the **Angle** edit box of the dialog box.
- Other parameters in the dialog box has been discussed earlier in the **Move/Copy** tool.
- After specifying desired parameters, click on the **OK** button from the dialog box. The entity will be rotated; refer to Figure-23.

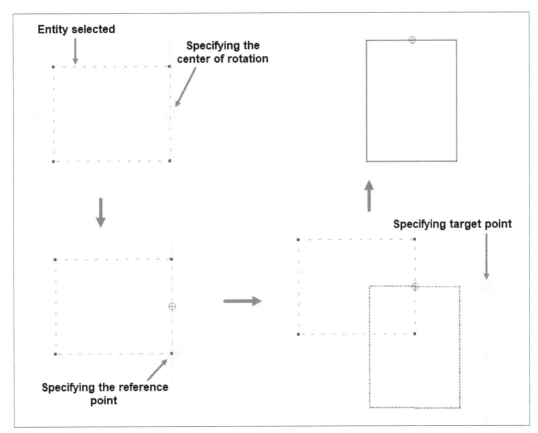

Figure-23. Entity rotated

- Press **Esc** button or click **RMB** to exit the tool.

Scale

Toolbar: Modify > Scale
Widget: Modify > Scale
Drop-Down: Modify > Scale
Menu: Tools > Modify > Scale

The **Scale** tool is used to increase or decrease the size of selected entity from a reference point by a defined factor for both axis. The procedure to use this tool is discussed next.

- Click on the **Scale** tool. You will be asked to select the entity to scale.
- Select the entity which you want to scale and press **Enter**. You will be asked to specify the reference point.
- Click in the Drawing Window at desired location to specify the reference point. The **Scaling Options** dialog box will be displayed; refer to Figure-24.

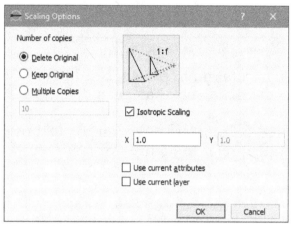

Figure-24. Scaling Options dialog box

- Specify desired value of x coordinate in the **X** edit box.
- Select the **Isotropic Scaling** check box and specify desired value of y coordinate in the **Y** edit box.
- Other options in the dialog box have been discussed earlier.
- After specifying desired parameters, click on the **OK** button from the dialog box. The entity will be scaled; refer to Figure-25.

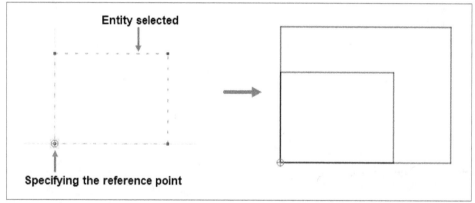

Figure-25. Entity scaled

- Press **Esc** button or click **RMB** to exit the tool.

Mirror

Toolbar: Modify > Mirror	
Widget: Modify > Mirror	
Drop-Down: Modify > Mirror	
Menu: Tools > Modify > Mirror	

The **Mirror** tool is used to create a mirror image of selected entity around an axis defined by two points. The procedure to use this tool is discussed next.

*   Click on the **Mirror** tool. You will be asked to select the entity to be mirrored.
*   Select desired entity in the Drawing Window which you want to mirror and press **Enter**. You will be asked to specify the first point of mirror line.
*   Click in the Drawing Window at desired location to specify the first point of mirror line. You will be asked to specify the second point and preview of mirror feature will be displayed.
*   Move the cursor away and click at desired location to specify the second point of mirror line. The **Mirroring Options** dialog box will be displayed; refer to Figure-26.

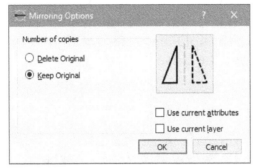

Figure-26. Mirroring Options dialog box

*   The parameters in the dialog box has been discussed earlier.
*   After specifying desired parameters, click on the **OK** button from the dialog box. The entity will be mirrored; refer to Figure-27.
*   Press **Esc** button or click **RMB** to exit the tool.

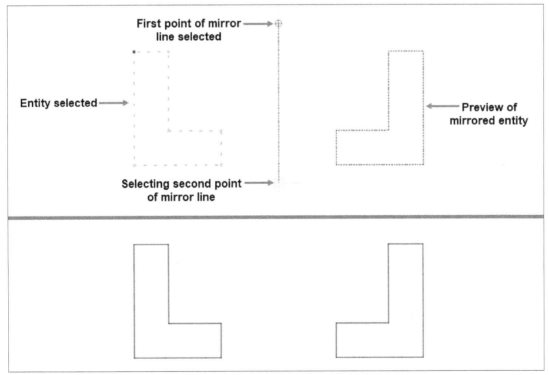

Figure-27. Mirrored entity created

Move and Rotate

Toolbar: Modify > Move and Rotate
Widget: Modify > Move and Rotate
Drop-Down: Modify > Move and Rotate
Menu: Tools > Modify > Move and Rotate

The **Move and Rotate** tool is used to move a selected entity by defining a reference point, a relative target point, and specifying the rotation angle. The procedure to use this tool is discussed next.

- Click on the **Move and Rotate** tool. You will be asked to select the entity to move and rotate.
- Select desired entity in the Drawing Window to move & rotate and press **Enter**. The **Tool Options** will be displayed; refer to Figure-28 and you will be asked to specify the reference point to move.

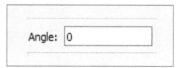

Figure-28. Move and Rotate Tool Options

- Specify desired angle of rotation in the **Angle** edit box.
- Click in the Drawing Window at desired location to specify the reference point to move the entity. The rotated entity will be attached to the cursor and you will be asked to specify the target point of entity.
- Move the cursor away and click at desired location to specify the target point of entity. The **Move/Rotate Options** dialog box will be displayed; refer to Figure-29.

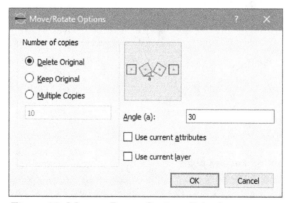

Figure-29. Move or Rotate Options dialog box

- All the parameters in the dialog box have been discussed earlier.
- After specifying desired parameters in the dialog box, click on the **OK** button. The entity will be rotated and moved; refer to Figure-30.

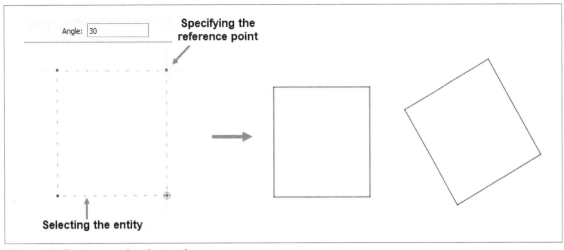

Figure-30. Entity moved and rotated

• Press **Esc** button or click **RMB** to exit the tool.

Rotate Two

Toolbar: Modify > Rotate Two
Widget: Modify > Rotate Two
Drop-Down: Modify > Rotate Two
Menu: Tools > Modify > Rotate Two

The **Rotate Two** tool is used to rotate selected entity about an absolute rotation point, while rotating the entity about a relative reference point to a target point. The procedure to use this tool is discussed next.

• Click on the **Rotate Two** tool. You will be asked to select the entity.
• Select desired entity to rotate and press **Enter**. You will be asked to specify absolute reference point.
• Click in the Drawing Window at desired location to specify the absolute reference point. You will be asked to specify relative reference point.
• Move the cursor away and click at desired location to specify the relative reference point. The **Rotate Two Options** dialog box will be displayed; refer to Figure-31.

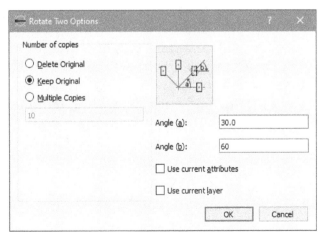

Figure-31. Rotate Two Options dialog box

• Specify desired angle value to rotate the entity in **Angle (a)** and **Angle (b)** edit boxes.

- Other parameters in the dialog box have been discussed earlier.
- After specifying desired parameters, click on the **OK** button from the dialog box. The entity will be rotated; refer to Figure-32.

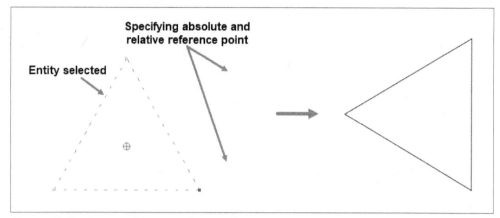

Figure-32. Entity rotated by two reference points

- Press **Esc** button or click **RMB** to exit the tool.

Revert direction

Toolbar: Modify > Revert direction
Widget: Modify > Revert direction
Drop-Down: Modify > Revert direction
Menu: Tools > Modify > Revert direction

The **Revert direction** tool is used to swap start and end points of one or more selected entities.

Trim

Toolbar: Modify > Trim
Widget: Modify > Trim
Drop-Down: Modify > Trim
Menu: Tools > Modify > Trim

The **Trim** tool is used to cut the length of an entity to an intersecting other entity. The procedure to use this tool is discussed next.

- Click on the **Trim** tool. You will be asked to select the limiting entity.
- Select desired entity in the Drawing Window to use as the limiting entity. You will be asked to select the entity(s) to trim.
- Select desired entity(s) to trim. The entity(s) will be trimmed; refer to Figure-33.

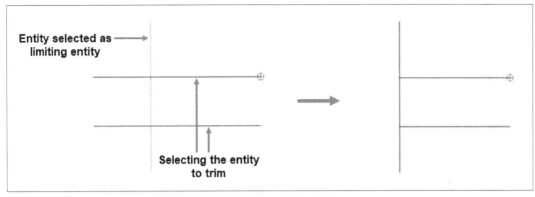

Figure-33. Entities trimmed

• Press **Esc** button or click **RMB** to exit the tool.

Trim Two

Toolbar: Modify > Trim Two
Widget: Modify > Trim Two
Drop-Down: Modify > Trim Two
Menu: Tools > Modify > Trim Two

The **Trim Two** tool is used to cut the lengths of two intersecting entities to the point of intersection. The procedure to use this tool is discussed next.

• Click on the **Trim Two** tool. You will be asked to select the first entity to trim.
• Select the first entity to be trimmed. You will be asked to select the second entity.
• Select the second entity intersecting with the first entity. The entity will be trimmed at the intersection of point; refer to Figure-34.

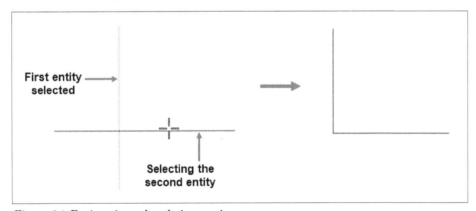

Figure-34. Entity trimmed at the intersection

• Press **Esc** button or click **RMB** to exit the tool.

Lengthen

Toolbar: Modify > Lengthen
Widget: Modify > Lengthen
Drop-Down: Modify > Lengthen
Menu: Tools > Modify > Lengthen

The **Lengthen** tool is used to extend or trim the length of an entity to another intersecting entity. The procedure to use this tool is discussed next.

- Click on the **Lengthen** tool. The **Tool Options** will be displayed; refer to Figure-35 and you will be asked to select the entity.

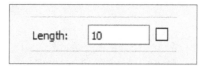

Figure-35. Lengthen Tool Options

- Specify desired value of extension length in the **Length** edit box.
- Select the check box to use the length specified in the **Length** edit box as total length after trimming, instead of length increase.
- Select desired entity to extend. The entity will be extended; refer to Figure-36.

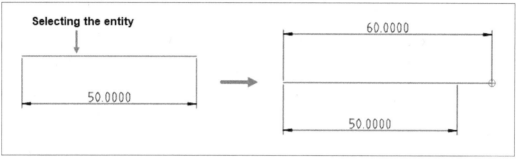

Figure-36. Entity extended

- Press **Esc** button or click **RMB** to exit the tool.

Offset

Toolbar: Modify > Offset	
Widget: Modify > Offset	
Drop-Down: Modify > Offset	
Menu: Tools > Modify > Offset	

The **Offset** tool is used to create copy of selected entity at defined distance in the specified direction. The procedure to use this tool is discussed next.

- Select desired object to create offset object and click on the **Offset** tool. The **Tool Options** will be displayed; refer to Figure-37 and the offset object will be attached to the cursor. Also, you will be asked to specify the direction of offset object.

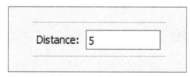

Figure-37. Offset Tool Options

- Specify desired offset distance in the **Distance** edit box.
- Specify desired direction of offset object in the Drawing Window and click **LMB**. The object will be offsetted; refer to Figure-38.

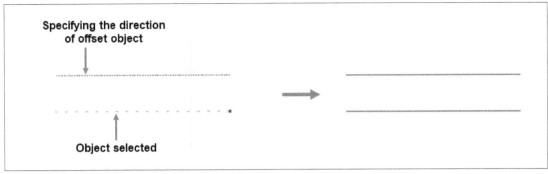

Figure-38. Object offsetted

Bevel

Toolbar: Modify > Bevel
Widget: Modify > Bevel
Drop-Down: Modify > Bevel
Menu: Tools > Modify > Bevel

The **Bevel** tool is used to create a chamfer between two line segments at the intersection point. The procedure to use this tool is discussed next.

- Click on the **Bevel** tool. The **Tool Options** will be displayed; refer to Figure-39 and you will be asked to select the first & then second entity.

Figure-39. Bevel Tool Options

- Select the **Trim** check box to trim both the entities to the bevel.
- Specify the length of first and second entity to bevel in the **Length 1** and **Length 2** edit boxes, respectively.
- Select the first entity and then the second entity to bevel. The entities will be bevelled; refer to Figure-40.

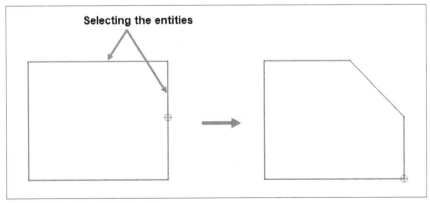

Figure-40. Entities bevelled

- Press **Esc** button or click **RMB** to exit the tool.

Fillet

Toolbar: Modify > Fillet
Widget: Modify > Fillet
Drop-Down: Modify > Fillet
Menu: Tools > Modify > Fillet

The **Fillet** tool is used to create a rounded edge at the intersection of two entities like line, arc, and so on with defined radius. The procedure to use this tool is discussed next.

- Click on the **Fillet** tool. The **Tool Options** will be displayed; refer to Figure-41 and you will be asked to select the first entity.

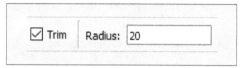

Figure-41. Fillet Tool Options

- Specify desired radius value for the fillet in the **Radius** edit box.
- Select the first entity. The filleted edge is attached to the cursor and you are asked to select the second entity.
- As you select the second entity, the filleted edge is applied to the entities; refer to Figure-42.

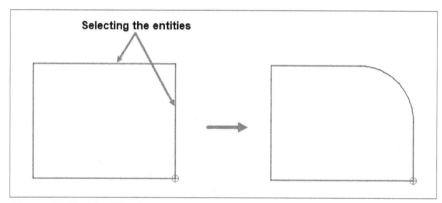

Figure-42. Entities filleted

- Press **Esc** button or click **RMB** to exit the tool.

Divide

Toolbar: Modify > Divide
Widget: Modify > Divide
Drop-Down: Modify > Divide
Menu: Tools > Modify > Divide

The **Divide** tool is used to divide or break a line at the selected "cutting" point. The procedure to use this tool is discussed next.

- Click on the **Divide** tool. You will be asked to select the entity to divide or break.
- Select desired entity to divide or break. You will be asked to specify the cutting point.

- Click on the entity at desired location to specify the cutting point. The entity will be divided; refer to Figure-43.

Figure-43. Entity divided

- Press **Esc** button or click **RMB** to exit the tool.

Stretch

Toolbar: Modify > Stretch
Widget: Modify > Stretch
Drop-Down: Modify > Stretch
Menu: Tools > Modify > Sretch

The **Stretch** tool is used to increase the size of a sketch section. The procedure to use this tool is discussed next.

- Click on the **Stretch** tool. You will be asked to select the first corner point.
- Click in the Drawing Window at desired location to specify the first corner point. You will be asked to specify the second corner point.
- Move the cursor away and click at desired location to specify the second corner point. You will be asked to specify the reference point.
- Move the cursor away and click at desired location to specify the reference point. You will be asked to specify the target point.
- Move the cursor away and click at desired location to specify the target point. The entity will be stretched; refer to Figure-44.

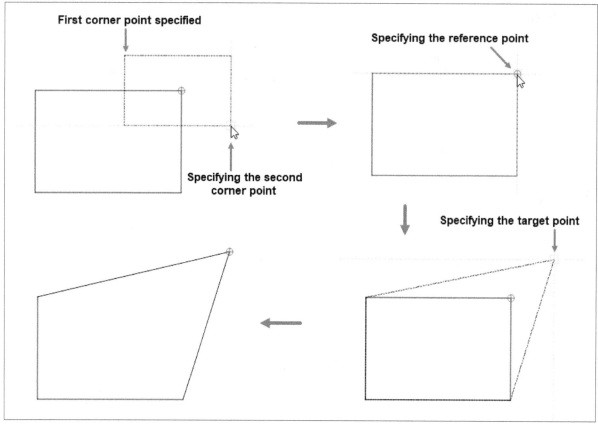

Figure-44. Entity stretched

- Press **Esc** button or click **RMB** to exit the tool.

Properties

Toolbar: Modify > Properties
Widget: Modify > Properties
Drop-Down: Modify > Properties
Menu: Tools > Modify > Properties

The **Properties** tool is used to modify the attributes of a single entity including layer, pen color/width/line type and the entity's geometry (varies by type of entity). The procedure to use this tool is discussed next.

- Click on the **Properties** tool. You will be asked to select the entity.
- Click on desired entity to modify the properties. The dialog box will be displayed as per the entity selected.
- In our case, the circle is selected. So, the **Circle** dialog box will be displayed; refer to Figure-45.

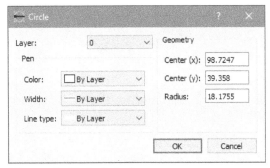

Figure–45. Circle dialog box

- Select desired layer from **Layer** drop-down.
- Select desired color of the entity from **Color** drop-down. Select **Custom** option from **Color** drop-down. The **Select Color** dialog box will be displayed; refer to Figure-46.

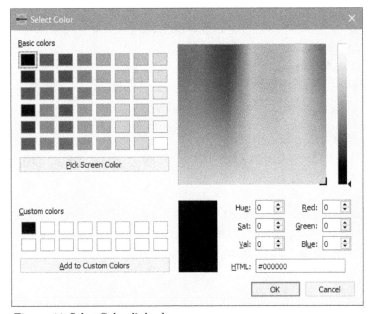

Figure–46. Select Color dialog box

- Select desired color from the **Select Color** dialog box and click on the **OK** button.
- Select desired width of entity from **Width** drop-down in the **Circle** dialog box.
- Select desired type of line from the **Line type** drop-down.
- Specify desired value for the x and y coordinates in the **Center (x)** and **Center (y)** edit boxes of the dialog box, respectively.
- Specify desired radius of the entity in the **Radius** edit box.
- After specifying desired parameters, click on the **OK** button from the dialog box. The entity will be modified; refer to Figure-47.

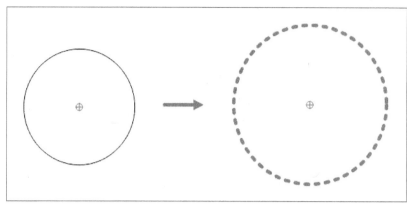

Figure–47. Entity modified

- Press **Esc** button or click **RMB** to exit the tool.

Attributes

Toolbar: Modify > Attributes
Widget: Modify > Attributes
Drop-Down: Modify > Attributes
Menu: Tools > Modify > Attributes

The **Attributes** tool is used to modify the common attributes of one or more selected entities including layer, pen color/width/line type. The procedure to use this tool is discussed next.

- Click on the **Attributes** tool. You will be asked to select the entity to modify.
- Select desired entity to modify the attributes and press **Enter**. The **Attributes** dialog box will be displayed; refer to Figure-48.

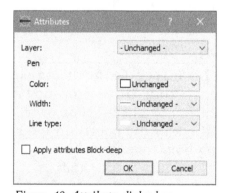

Figure–48. Attributes dialog box

- Select **Apply attributes Block-deep** check box to apply attributes also to all sub-entities of selected object. This recursively modifies all entities of the Block itself.
- Other parameters in the dialog box has been discussed earlier.
- After specifying desired parameters, click on the **OK** button from the dialog box. The entity will be modified; refer to Figure-49.

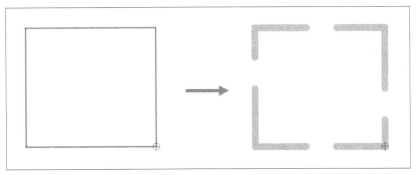

Figure-49. Entity modified

Explode Text into Letters

Toolbar: Modify > Explode Text into Letters
Widget: Modify > Explode Text into Letters
Drop-Down: Modify > Explode Text into Letters
Menu: Tools > Modify > Explode Text into Letters

The **Explode Text into Letters** tool is used to separate a string of text into individual character entities. The procedure to use this tool is discussed next.

- Click on the **Explode Text into Letters** tool. You will be asked to select the string of text.
- Select desired string of text in the Drawing Window to explode into individual letters and press **Enter**. The string of text will be exploded into individual letters; refer to Figure-50.

OR,

- Select desired string of text in the Drawing Window to explode and click on the **Explode Text into Letters** tool. The text will be exploded.

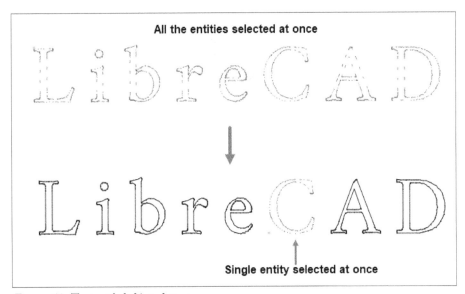

Figure-50. Text exploded into letters

Explode

Toolbar: Modify > Explode
Widget: Modify > Explode
Drop-Down: Modify > Explode
Menu: Tools > Modify > Explode

The **Explode** tool is used to separate one or more selected blocks or compound entities into individual entities. The procedure to use this tool is discussed next.

- Click on the **Explode** tool. You will be asked to select the compound entities.
- Select desired compound entities to explode into individual entities and press **Enter**. The entities will be exploded; refer to Figure-51.

OR

- Select desired compound entities to explode and click on the **Explode** tool. The entities will be exploded.

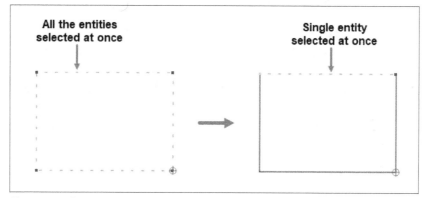

Figure-51. Compound entities exploded into individual entities

Delete selected

Toolbar: Modify > Delete selected
Widget: Modify > Delete selected
Drop-Down: Modify > Delete selected
Menu: Tools > Modify > Delete selected

The **Delete selected** tool is used to delete one or more selected entities. The procedure to use this tool is discussed next.

- Select desired entity(s) in the Drawing Window to delete and click on the **Delete selected** tool. The selected entity will be deleted; refer to Figure-52.

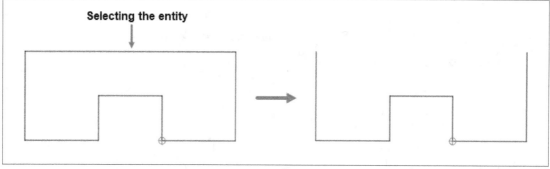

Figure-52. Selected entity deleted

INFO TOOLS

Info tools are available in the **Info** cascading menu of the **Tools** menu; refer to Figure-53, **Info** widget; refer to Figure-54, and **Info** toolbar; refer to Figure-55. These tools are discussed next.

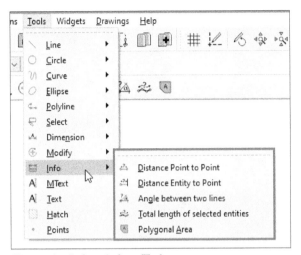

Figure-54. Info widget

Figure-53. Info tools from Tools menu

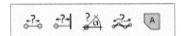

Figure-55. Info toolbar

Distance Point to Point

Toolbar: Info > Distance Point to Point
Widget: Info > Distance Point to Point
Menu: Tools > Info > Distance Point to Point

The **Distance Point to Point** tool is used to display distance between two points along with coordinates of points in Cartesian as well as Polar System. The procedure to use this tool is discussed next.

- Click on the **Distance Point to Point** tool. You will be asked to specify the first point.
- Click on desired entity for the first point to provide the distance. You will be asked to specify the second point.
- Move the cursor away and click on the entity for the second point. The distance, cartesian, and polar coordinates will be displayed in the **Command history**; refer to Figure-56.

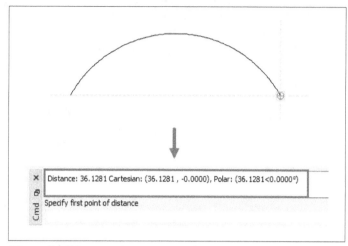

Figure-56. Distance point to point tool

Distance Entity to Point

Toolbar: Info > Distance Entity to Point
Widget: Info > Distance Entity to Point
Menu: Tools > Info > Distance Entity to Point

The **Distance Entity to Point** tool is used to display shortest distance between selected entity and specified point. The procedure to use this tool is discussed next.

- Click on the **Distance Entity to Point** tool. You will be asked to select the entity.
- Select desired entity in the Drawing Window from which the distance is to be measured. You will be asked to select the point.
- Move the cursor away and click on the other entity at desired location to specify the point upto which the distance is to be measured. The shortest distance will be displayed in the **Command history**; refer to Figure-57.

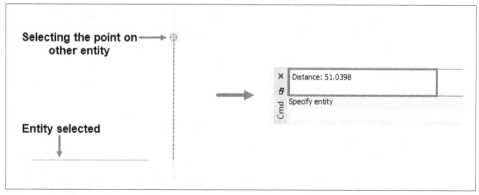

Figure-57. Shortest distance displayed

- Press **Esc** button or click **RMB** to exit the tool.

Angle between two lines

Toolbar: Info > Angle between two lines
Widget: Info > Angle between two lines
Menu: Tools > Info > Angle between two lines

The **Angle between two lines** tool is used to provide angle between two selected line segments, measured counter-clockwise. The procedure to use this tool is discussed next.

- Click on the **Angle between two lines** tool. You will be asked to select the first line and then the second line.
- Select first line in the Drawing Window and then the second line between which you want to measure the angle. The angle between the lines will be displayed in the **Command history**; refer to Figure-58.

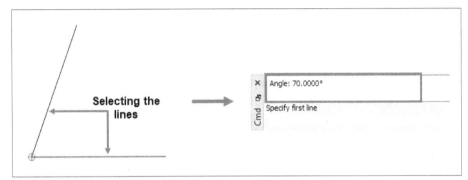

Figure-58. Angles between the lines displayed

- Press **Esc** button or click **RMB** to exit the tool.

Total length of selected entities

Toolbar: Info > Total length of selected entities
Widget: Info > Total length of selected entities
Menu: Tools > Info > Total length of selected entities

The **Total length of selected entities** tool is used to provide total length of one or more selected entities (length of line segment, circle circumference, etc.) The procedure to use this tool is discussed next.

- Click on the **Total length of selected entities** tool. You will be asked to select the entity(s).
- Select desired entity(s) in the Drawing Window to measure the total length and press **Enter**. The total length of the entity(s) will be displayed in the **Command history**; refer to Figure-59.

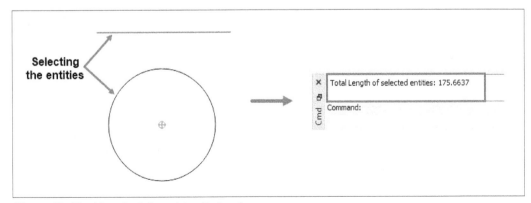

Figure-59. Total length of the entities displayed

Polygonal Area

Toolbar: Info > Polygonal Area
Widget: Info > Polygonal Area
Menu: Tools > Info > Polygonal Area

The **Polygonal Area** tool is used to display area and circumference of polygon defined by three or more specified points. The procedure to use this tool is discussed next.

*   Click on the **Polygonal Area** tool. You will be asked to specify the first point.
*   Click in the Drawing Window at desired location to specify the first point of polygon. You will be asked to specify the next point.
*   Move the cursor away and click at desired location to specify the next point of polygon. You will be asked to specify the next point(s).
*   Move the cursor away and click at desired location to specify the next point(s) of polygon. The polygon will be created and the area of the polygon will be displayed in the **Command history**; refer to Figure-60.

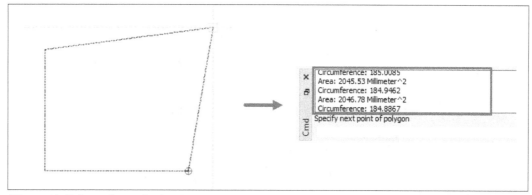

Figure-60. Polygonal area displayed

MISCELLANEOUS TOOLS

Other tools are available at the bottom in the **Tools** menu, refer to Figure-61 and **DefaultCustom** toolbar; refer to Figure-62. These tools are discussed next.

Figure-62. DefaultCustom toolbar

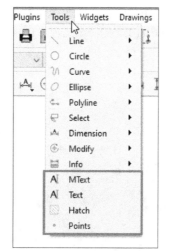

Figure-61. Other tools in Tools menu

MText

Toolbar: DefaultCustom > Polygonal Area
Menu: Tools > Polygonal Area

The **MText** tool is used to insert multi-line text into drawing at a specified base point. The procedure to use this tool is discussed next.

- Click on the **MText** tool. The **MText** dialog box will be displayed; refer to Figure-63.

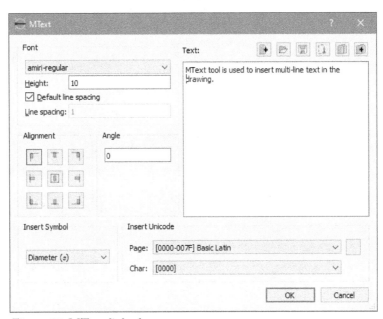

Figure-63. MText dialog box

- Select desired font from the drop-down in the **Font** area of the dialog box.
- Specify desired height of the text in the **Height** edit box.
- Select the **Default line spacing** check box to specify the distance between the lines by default.
- Specify desired spacing of the line in the **Line spacing** edit box.
- Select desired button from the **Alignment** area of the dialog box to align the text.
- Specify desired angle of the text in the **Angle** edit box.
- Select desired symbol to insert in the text from the drop-down in the **Insert Symbol** area of the dialog box.
- Select desired unicode symbols from **Page** and **Char** edit boxes from **Insert Unicode** area of the dialog box. Click on the ▢ button next to the **Page** drop-down to insert the unicode selected in the **Page** and **Char** drop-downs in the **Text** box.
- Click on the **Clear Text** ▣ button to clear the text specified in the **Text** box.
- Click on the **Load Text From File** ▣ button to load the text from the local drive. The **Open** dialog box will be displayed; refer to Figure-64. Select desired text file to load and click on the **Open** button from the dialog box. The text file will be opened.

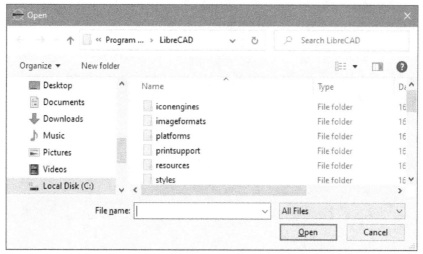

Figure-64. Open dialog box

- Click on the **Save Text To File** button to save the text. The **Save As** dialog box will be displayed; refer to Figure-65. Specify desired name for the text file in the **File name** edit box and click on the **Save** button. The text file will be saved.

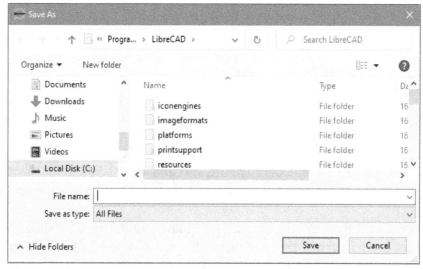

Figure-65. Save As dialog box

- Select the text from the **Text** box and click on the **Cut** button. The text will be cut.
- Select the text from the **Text** box and click on the **Copy** button. The text will be copied.
- Click in the **Text** box and and then click on the **Paste** button to paste the copied or cut text.
- After specifying desired parameters, click on the **OK** button from the dialog box. The **Tool Options** will be displayed along with the box attached to cursor; refer to Figure-66.

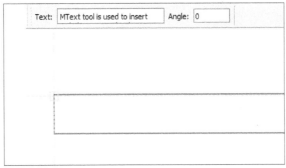

Figure-66. Mtext Tool Options and box attached to cursor

- Modify the text and angle of text as desired in the **Text** edit box and **Angle** edit box, respectively.
- Click in the Drawing Window at desired location to place the text. The text will be placed; refer to Figure-67.

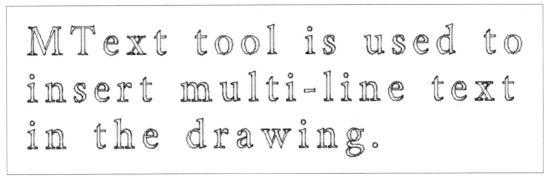

Figure-67. MText created

- Move the cursor away and click to place the text at another location(s) or press **Esc** button or click **RMB** to exit the tool.

Text

Menu: Tools > Text

The **Text** tool is used to insert single line text into drawing at a specified base point. The procedure to use this tool is discussed next.

- Click on the **Text** tool. The **Text** dialog box will be displayed; refer to Figure-68.

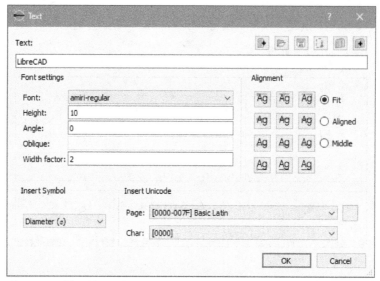

Figure-68. Text dialog box

- Specify desired text in the **Text** edit box.
- Select desired font of text from **Font** drop-down in the **Font settings** area of the dialog box.
- Specify height of the text as desired in the **Height** edit box.
- Specify angle of text as desired in the **Angle** edit box.
- Specify desired width of text in the **Width factor** edit box.
- Other parameters in the dialog box are same as discussed in **MText** dialog box and the procedure to use this tool is same as discussed for **MText** tool.

Hatch

Toolbar: DefaultCustom > Hatch
Menu: Tools > Hatch

The **Hatch** tool is used to fill a closed entity (polygon, circle, polyline, etc.) with a defined pattern or a solid fill. The procedure to use this tool is discussed next.

- Click on the **Hatch** tool. You will be asked to select the closed entity.
- Select desired closed entity from the Drawing Window. The **Choose Hatch Attributes** dialog box will be displayed; refer to Figure-69.
  OR,
- Select desired close entity and click on the **Hatch** tool. The **Choose Hatch Attributes** dialog box will be displayed.

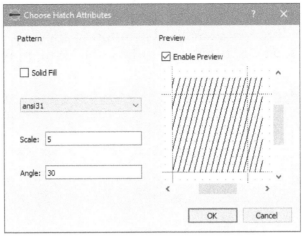

Figure-69. Choose Hatch Attributes dialog box

- Select the **Solid Fill** check box from **Pattern** area of the dialog box to fill the entity with solid fill.
- Select desired type of hatch pattern from the drop-down in the **Pattern** area.
- Specify desired value for scaling the pattern in the **Scale** edit box.
- Specify desired angle for the text in the **Angle** edit box.
- Select **Enable Preview** check box from **Preview** area of the dialog box to display the preview of hatch pattern selected.
- After specifying desired parameters, click on the **OK** button from the dialog box. The hatch pattern will be applied to the entity; refer to Figure-70.

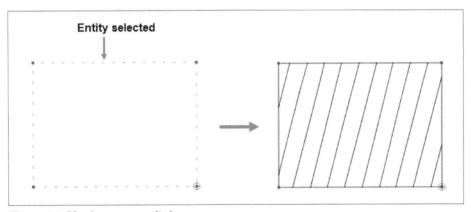

Figure-70. Hatch pattern applied

Insert Image

Toolbar: DefaultCustom > Insert Image

The **Insert Image** tool is used to insert an image from the local drive. The procedure to use this tool has been discussed in Chapter 1.

Create Block

The **Create Block** tool is used to create block. The procedure to use this tool is discussed next.

- Click on the **Create Block** tool. You will be asked to select the entity.

- Select desired entities to be included in creating the block and press **Enter**. You will be asked to specify the reference point.
- Click in the Drawing Window at desired location to specify the reference point; refer to Figure-71. The **Block Settings** dialog box will be displayed; refer to Figure-72.

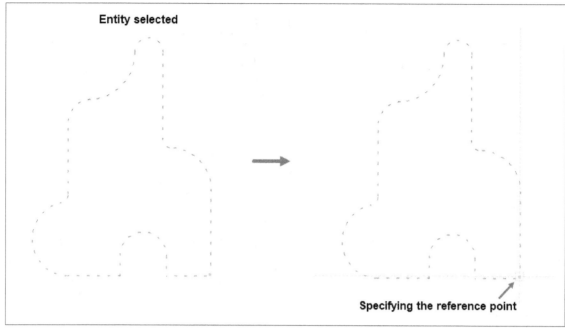

Figure-71. Specifying the reference point

Figure-72. Block Settings dialog box

- Specify desired name of the block in the **Block Name** edit box and click on the **OK** button from the dialog box. The block will be created and the block name will be displayed in the **Block List** toolbar; refer to Figure-73.

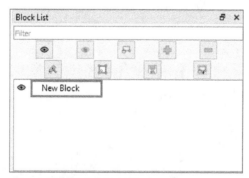

Figure-73. Block List toolbar

- Click on the **Show all blocks** or **Hide all blocks** button to show or hide all the created blocks, respectively.
- Click on the **Create Block** button to create a new block.
- Click on the **Add an empty block** button. The **Block Settings** dialog box will be displayed as discussed earlier. Specify the parameters as desired.

- Click on the **Remove block** ═ button. A warning message box will be displayed to remove the blocks; refer to Figure-74. Click on the **OK** button to remove the blocks.

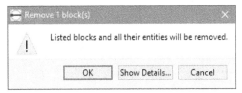

Figure-74. Warning message box

- Click on the **Rename the active block** ⚒ button. The **Block Settings** dialog box will be displayed as discussed earlier. Specify the parameters as desired.
- Click on the **Edit the active block in a separate window** ☒ button. The existing block to be edited will be displayed in the Drawing Window.
- Click on the **save the active block to a file** 🖫 button. The **Save Block As** dialog box will be displayed; refer to Figure-75. Specify desired name of the block in the **File name** edit box and click on the **Save** button. The block will be saved.

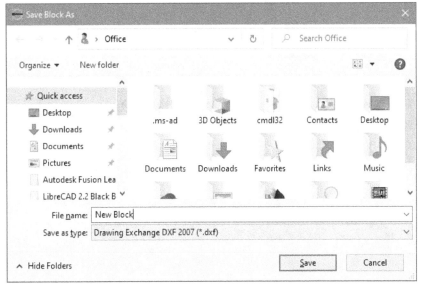

Figure-75. Save Block As dialog box

- Click on **Insert the active block** 🖫 button to insert the active existing block in the Drawing Window.

Points

The **Points** tool is used to draw a point at the assigned coordinates. The procedure to use this tool is discussed next.

- Click on the **Points** tool. You will be asked to specify the location of point.
- Click in the Drawing Window at desired location(s) to specify the point(s) or you can input coordinates in Command Line. The point(s) will be created; refer to Figure-76.

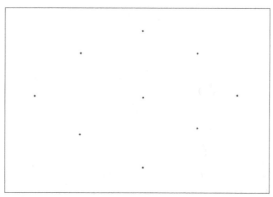
Figure-76. Points created

SNAP SELECTION TOOLS

Snapping tools are available in the **Snap Selection** toolbar, refer to Figure-77. These tools are discussed next.

Figure-77. Snap Selection toolbar

Exclusive Snap Mode ‹Ex›

When **Exclusive Snap Mode (Ex)** button is selected then only one snap mode is allowed. If **Ex** button is not selected then multiple snap modes are allowed; refer to Figure-78.

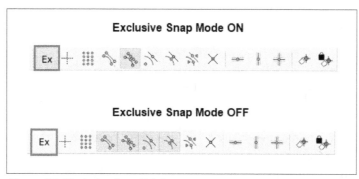
Figure-78. Exclusive Snap Mode

Free Snap

The ⊹ **Free Snap** button allow for the crosshair to move freely while other snap modes are enabled.

Snap on Grid

The ⠿ **Snap on Grid** button snap to a grid intersection.

Snap on Endpoints

The ⌐ **Snap on Endpoints** button snap to the endpoints of a line segment, the quadrants of a circle, a point, or the alignment point of a text or mtext object.

Snap on Entity

The [icon] **Snap on Entity** button snaps to the path of an entity.

Snap Center

The [icon] **Snap Center** button snaps to the center of a circle or ellipse. It will also snap to the foci of an ellipse.

Snap Middle

The [icon] **Snap Middle** button snaps to the middle of a path. Enabling this mode displays a "Middle points" input. If you change the value to 2 then you can snap to the trisection points of a line segment.

Snap Distance

If you snap to the endpoint of a line segment then activate [icon] "**Snap Distance**" and input **50** in the **Distance** edit box displayed in the **Tool Options**, then it will snap to a point **50** units from the endpoint on the line segment. However, it will also snap to a point that is **50** units from the other point.

Snap Intersection

The [icon] **Snap Intersection** button snaps to the intersection of two entities. Note that this does not currently work for polylines.

Restrict Horizontal

The [icon] **Restrict Horizontal** button restricts the crosshairs to the x-axis (horizontal movement).

Restrict Vertical

The [icon] **Restrict Vertical** button restricts the crosshairs to the y-axis (vertical movement).

Restrict Orthogonal

The [icon] **Restrict Orthogonal** button restricts the crosshairs to the x or y-axis. (either horizontal or vertical movement).

Set relative zero position

The [icon] **Set relative zero position** button manually sets the relative zero point at the selected coordinate.

Lock relative zero position

The [icon] **Lock relative zero position** button locks the relative zero point to the current coordinate.

SELF ASSESSMENT

Q1. Which of the following tools is used to modify the attributes of a single entity including layer, pen color/width/line type and the entity's geometry (varies by type of entity)?

(a) Attributes
(b) Fillet
(c) Properties
(d) None of the Above

Q2. Which of the following tools is used to provide distance along with cartesian and polar coordinates of two specified points?

(a) Distance Entity to Point
(b) Distance Point to Point
(c) Total length of selected entities
(d) None of the Above

Q3. The _____ tool is used to apply dimension lines and text aligned to an existing entity.

Q4. The _____ tool is used to move selected entity by defining a reference point and a relative target point.

Q5. The **Rotate** tool is used to rotate selected entity about a _____ point, moving the entity from a reference point to a target point.

Q6. The _____ tool is used to increase or decrease the size of selected entity with respect to a reference point by defined factor for both axis.

Q7. The **Move and Rotate** tool is used to rotate selected entity by defining a reference point and a relative target point and moving the entity at a given angle. (True/False)

Q8. The **Offset** tool is used to create copy of selected entity at a defined distance in the specified direction. (True/False)

Q9. The **Fillet** tool is used to create chamfer between two intersecting line segments with defined by a setback on each segment. (True/False)

Ans: **1.** (c) **2.** (b) **3.** Aligned **4.** Move/Copy **5.** rotation **6.** Scale **7.** False **8.** True **9.** False

Chapter 4

Practical and Practices

The major topics covered in this chapter are:

- *Practical*
- *Practices*

PRACTICAL 1

Create the sketch as shown in Figure-1. Also, dimension the sketch as per the figure. Note that all the dimensions are in **inch**.

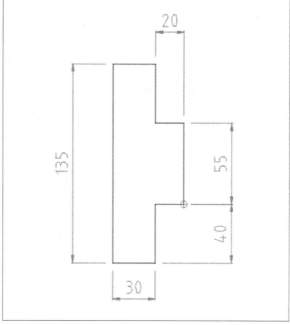

Figure-1. Practical 1

Creating Sketch

- Start **LibreCAD** if not started yet.
- Click on the **Vertical** tool from **Lines** drop-down. The **Tool Options** will be displayed and a line will be attached to the cursor.
- Specify the length of first vertical line as **135** in the **Length** edit box and select **Start** option from **Snap Point** drop-down of **Tool Options**.
- Click in the Drawing Window at desired location to create the vertical line. The first vertical line will be created; refer to Figure-2.

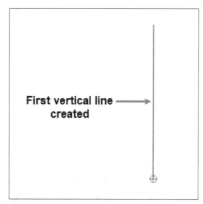

First vertical line ———▶
created

Figure-2. First vertical line created

- Press **Esc** button or click **RMB** to exit the tool.
- Click on the **Horizontal** tool from **Lines** drop-down. The **Tool Options** will be displayed and a horizontal line will be attached to the cursor.
- Specify the length of line as **30** in the **Length** edit box and select **Start** option from **Snap Point** drop-down of **Tool Options**.

- Select **Snap on Entity** button from **Snap Selection** toolbar and click at the endpoint of the **135** vertical line. The first horizontal line will be created; refer to Figure-3.

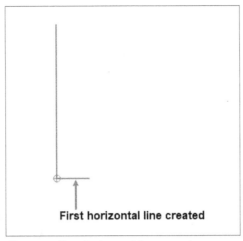

Figure-3. First horizontal line created

- Press **Esc** button or click **RMB** to exit the tool.
- Click on the **Vertical** tool from **Lines** drop-down. The **Tool Options** will be displayed and a vertical line will be attached to the cursor.
- Specify the length of line as **40** in the **Length** edit box and select **Start** option from **Snap Point** drop-down of **Tool Options**.
- Click at the endpoint of **30** horizontal line. The second vertical line will be created; refer to Figure-4.

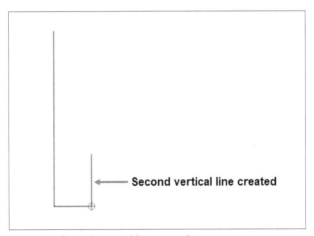

Figure-4. Second vertical line created

- Press **Esc** button or click **RMB** to exit the tool.
- Click on the **Horizontal** tool from **Lines** drop-down. The **Tool Options** will be displayed and a horizontal line will be attached to cursor.
- Specify the length of line as **20** in the **Length** edit box and select **Start** option from **Snap Point** drop-down of **Tool Options**.
- Click at the endpoint of **30** vertical line. The second horizontal line will be created; refer to Figure-5.

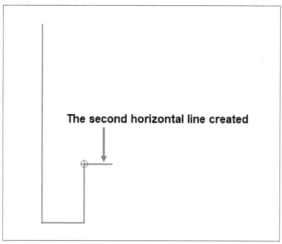

Figure-5. Second horizontal line created

- Press **Esc** button or click **RMB** to exit the tool.
- Click on the **Vertical** tool from **Lines** drop-down. The **Tool Options** will be displayed and a vertical line will be attached to the cursor.
- Specify the length of line as **55** in the **Length** edit box and select **Start** option from **Snap Point** drop-down of **Tool Options**.
- Click at the endpoint of **20** horizontal line. The third vertical line will be created; refer to Figure-6.

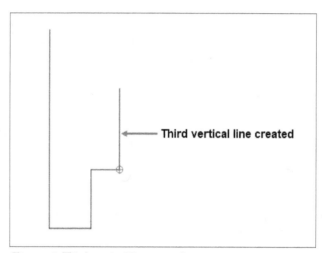

Figure-6. Third vertical line created

- Press **Esc** button or click **RMB** to exit the tool.
- Click on the **Horizontal** tool from **Lines** drop-down. The **Tool Options** will be displayed and a horizontal line will be attached to cursor.
- Specify the length of line as **20** in the **Length** edit box and select **End** option from **Snap Point** drop-down of **Tool Options**.
- Click at the endpoint of **55** vertical line. The third horizontal line will be created; refer to Figure-7.

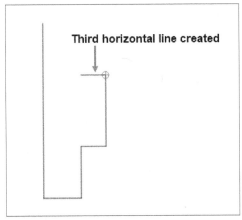

Figure-7. Third horizontal line created

- Press **Esc** button or click **RMB** to exit the tool.
- Click on the **Vertical** tool from **Lines** drop-down. The **Tool Options** will be displayed and a vertical line will be attached to the cursor.
- Specify the length of line as **40** in the **Length** edit box and select **Start** option from **Snap Point** drop-down.
- Click at the endpoint of **20** horizontal line. The fourth vertical line will be created; refer to Figure-8.

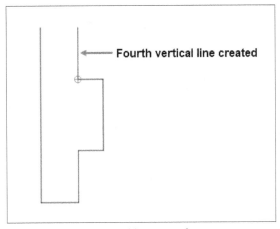

Figure-8. Fourth vertical line created

- Press **Esc** button or click **RMB** to exit the tool.
- Click on the **Horizontal** tool from **Lines** drop-down. The **Tool Options** will be displayed and a horizontal line will be attached to the cursor.
- Specify the length of line as **30** in the **Length** edit box and select **End** option from **Snap Point** drop-down.
- Click at the endpoint of **40** vertical line. The fourth horizontal line will be created; refer to Figure-9.

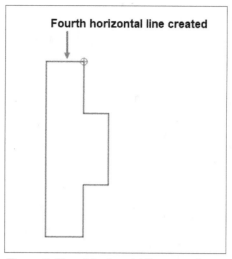

Figure-9. Fourth horizontal line created

- Press **Esc** button or click **RMB** to exit the tool.

Applying Dimensions

- Click on the **Aligned** tool from **Dimensions** drop-down and then select **Snap on Entity** button from **Snap Selection** toolbar.
- Select the endpoints of first vertical line. The dimension will be attached to cursor.
- Select **Free Snap** button from **Snap Selection** toolbar.
- Move the cursor at desired location and click to apply the dimension. The dimension will be applied; refer to Figure-10.

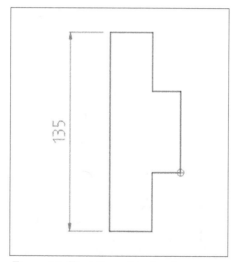

Figure-10. First dimension placed

- Similarly, apply all the dimensions of sketch. The sketch will be completed; refer to Figure-11.

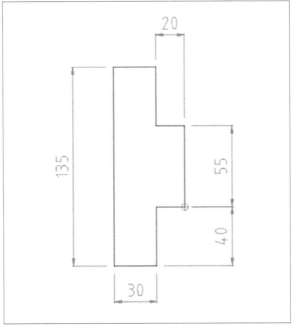

Figure-11. Practical 1 created

- Press **Esc** button or click **RMB** to exit the tool.

PRACTICAL 2

In this practical, we will create the sketch as shown in Figure-12. Note that all the dimensions are in **inch**.

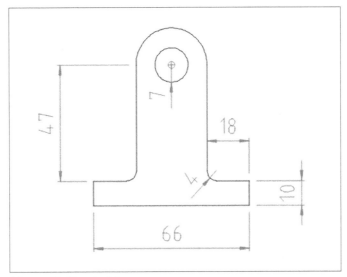

Figure-12. Practical 2

Starting a New Sketch

- Start LibreCAD if not started yet.
- Click on the **Horizontal** tool from **Lines** drop-down. The **Tool Options** will be displayed and a horizontal line will be attached to the cursor.
- Specify the length of line as **66** in the **Length** edit box of **Tool Options**.
- Click in the Drawing Window at desired location. The first horizontal line will be created; refer to Figure-13.

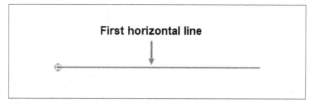

Figure-13. First horizontal line created

- Press **Esc** button or click **RMB** to exit the tool.
- Click on the **Vertical** tool from **Lines** drop-down. The **Tool Options** will be displayed and a vertical line will be attached to cursor.
- Specify the length of line as **10** in the **Length** edit box and select **Start** option from **Snap Point** drop-down of **Tool Options**.
- Select **Snap on Entity** button from **Snap Selection** toolbar.
- Click on the endpoint of **66** horizontal line. The first vertical line will be created; refer to Figure-14.

Figure-14. First vertical line created

- Similarly, create the second vertical line at the other endpoint of the **66** horizontal line; refer to Figure-15.

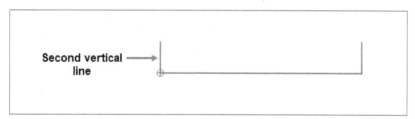

Figure-15. Second vertical line created

- Press **Esc** button or click **RMB** to exit the tool.
- Click on the **Horizontal** tool from **Lines** drop-down. The **Tool Options** will be displayed and a horizontal line will be attached to cursor.
- Specify the lenght of line as **18** in the **Length** edit box and select **End** option from **Snap Point** drop-down.
- Click at the endpoint of first vertical line. The second horizontal line will be created.
- Select **Start** option from **Snap Point** drop-down and click at the endpoint of second vertical line. The third horizontal line will be created; refer to Figure-16.

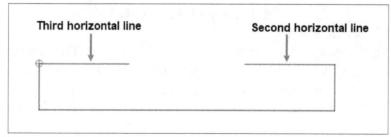

Figure-16. Second and third horizontal lines created

- Press **Esc** button or click **RMB** to exit the tool.
- Click on the **Vertical** tool from **Lines** drop-down. The **Tool Options** will be displayed and a vertical line will be attached to cursor.
- Specify the length of line as **47** in the **Length** edit box and select **Start** option from **Snap Point** drop-down of **Tool Options**.
- Click at the endpoint of second and third horizontal lines. The third and fourth vertical lines will be created; refer to Figure-17.

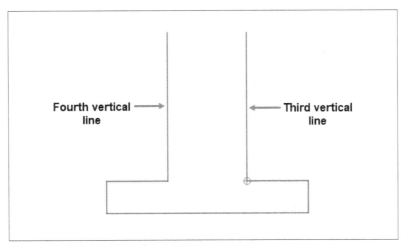

Figure-17. Third and fourth vertical lines created

- Press **Esc** button or click **RMB** to exit the tool.
- Click on the **2 Points** tool from **Circles** drop-down.
- Click at the endpoints of third and fourth vertical line. The circle will be created; refer to Figure-18.

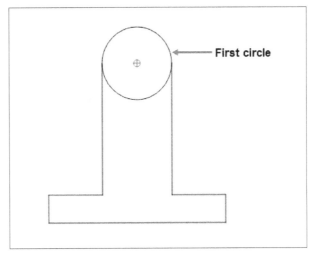

Figure-18. First circle created

- Press **Esc** button or click **RMB** to exit the tool.
- Click on the **Center, Radius** tool from **Circles** drop-down. The **Tool Options** will be displayed.
- Specify the radius of circle as **6.5** in the **Radius** edit box.
- Select **Snap Center** option from **Snap Selection** toolbar.
- Click at the center point of first circle. The second circle will be created; refer to Figure-19.

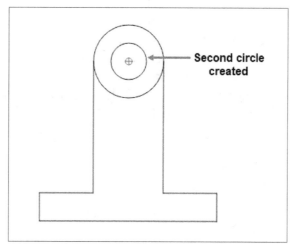

Figure-19. Second circle created

- Press **Esc** button or click **RMB** to exit the tool.
- Click on the **Fillet** tool from **Modify** drop-down. The **Tool Options** will be displayed.
- Specify the fillet radius as **4** in the **Radius** edit box.
- Click on the second horizontal and third vertical line. The fillet will be applied.
- Similarly, click on the third horizontal and fourth vertical line. The fillet will be applied; refer to Figure-20.

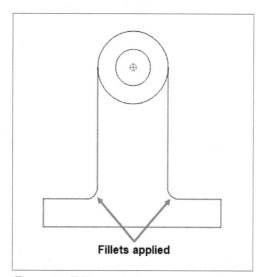

Figure-20. Fillets applied

- Press **Esc** button or click **RMB** to exit the tool.
- Click on the **2 Points** tool from **Lines** drop-down. The **Tool Options** will be displayed.
- Select **Snap on Endpoints** button from **Snap Selection** toolbar.
- Click at the endpoints of third and fourth vertical lines. The fourth horizontal line will be created; refer to Figure-21.

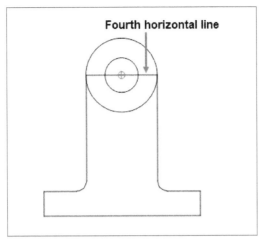

Figure-21. Fourth horizontal line created

- Click on the **Trim** tool from **Modify** drop-down.
- Select the fourth horizontal line as the limiting entity and click on the entity to trim. The entity will be trimmed; refer to Figure-22.

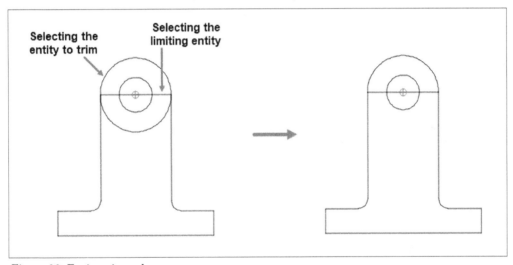

Figure-22. Entity trimmed

- Press **Esc** button or click **RMB** to exit the tool.
- Select the fourth horizontal line and click on the **Delete** button from the keyboard. The sketch will be created; refer to Figure-23.

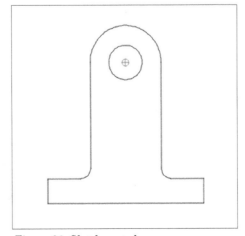

Figure-23. Sketch created

Applying Dimensions

- Click on the **Horizontal** tool from **Dimensions** drop-down and select **Snap on Endpoints** button from **Snap Selection** toolbar.
- Click at the endpoints of first horizontal line. The dimension will be attached to cursor.
- Select **Free Snap** button from **Snap Selection** toolbar.
- Move the cursor at desired location and click to apply the dimension. The first dimension will be applied; refer to Figure-24.

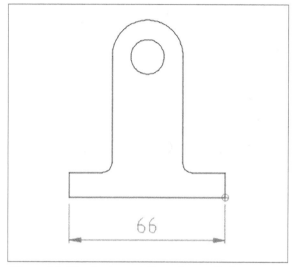

Figure-24. First horizontal dimension applied

- Similarly, apply the dimension to second horizontal line; refer to Figure-25.

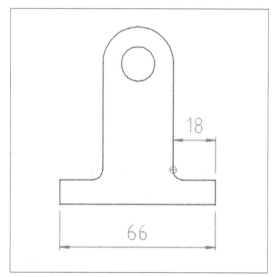

Figure-25. Second horizontal dimension applied

- Press **Esc** button or click **RMB** to exit the tool.
- Click on the **Vertical** tool from **Dimensions** drop-down.
- Click at the endpoints of first vertical line. The first vertical dimension will be applied; refer to Figure-26.

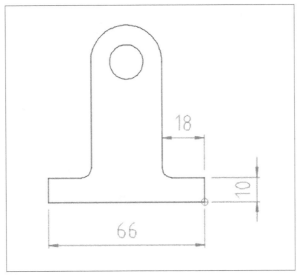

Figure-26. First vertical dimension applied

- Similarly, apply the dimension to fourth vertical line; refer to Figure-27.

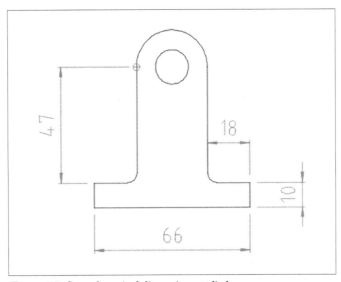

Figure-27. Second vertical dimension applied

- Press **Esc** button or click **RMB** to exit the tool.
- Click on the **Radial** tool from **Dimensions** drop-down and click at the first created fillet. The radial dimension will be applied; refer to Figure-28.

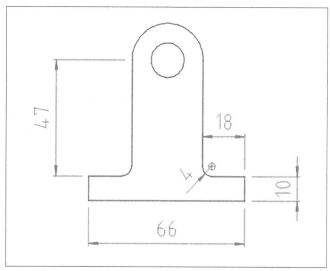

Figure-28. Radial dimension applied

- Similarly, apply the dimension to second created circle. The sketch will be created along with the dimensions; refer to Figure-29.

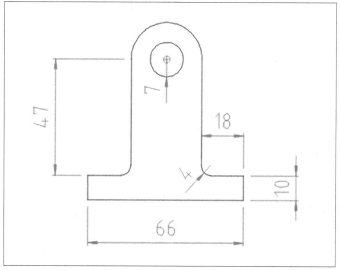

Figure-29. Practical 2 created

- Press **Esc** button or click **RMB** to exit the tool.

PRACTICAL 3

In this practical, we will create the sketch as shown in Figure-30. Note that all the dimensions are in **inch**.

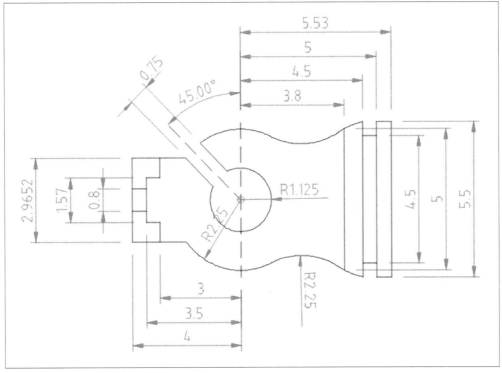

Figure-30. Practical 3

Starting a New Sketch

- Start LibreCAD if not started yet.
- Select **Center, Radius** tool from **Circle** toolbar. The **Tool Options** will be displayed and a circle will be attached to cursor.
- Specify the radius of circle as **1.125** in the **Radius** edit box.
- Click in the Drawing Window at desired location. The first circle will be created; refer to Figure-31.

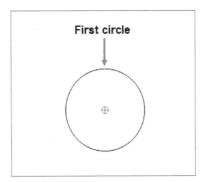

Figure-31. First circle created

- Similarly, specify radius of second circle as **2.50** in the **Radius** edit box and click at the center point of first circle. The second circle will be created; refer to Figure-32.

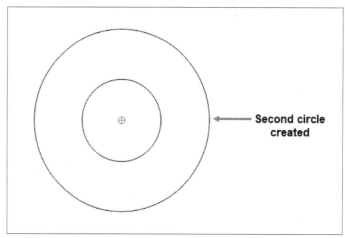

Figure-32. Second circle created

- Press **Esc** button or click **RMB** to exit the tool.
- Click on the **Vertical** tool from **Line** toolbar. The **Tool Options** will be displayed and a line will be attached to cursor.
- Specify the length of line as **7** in the **Length** edit box and select **Middle** option from **Snap Point** drop-down.
- Select the **Dash (small)** option from the **Line type** drop-down.
- Click at the center point of circles. The first dashed line will be created; refer to Figure-33.

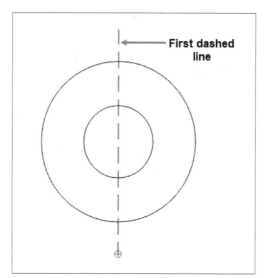

Figure-33. First dashed line created

- Press **Esc** button or click **RMB** to exit the tool.
- Click on the **Angle** tool from **Line** toolbar. The **Tool Options** will be displayed and an angled line will be attached to cursor.
- Specify the angle of line as **135** in the **Angle** edit box, length of line as **4** in the **Length** edit box, and select **Start** option from **Snap Point** drop-down.
- Click at the center point of circles. The second dashed line will be created; refer to Figure-34.

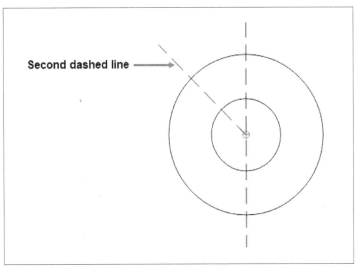

Figure-34. Second dashed line created

- Press **Esc** button or click **RMB** to exit the tool.
- Select second dashed line and click on the **Offset** tool from **Modify** toolbar. The **Tool Options** will be displayed and an offset line will be attached to cursor.
- Specify the offset distance as **0.375** in the **Distance** edit box and create offset lines at both the sides of second dashed line; refer to Figure-35.

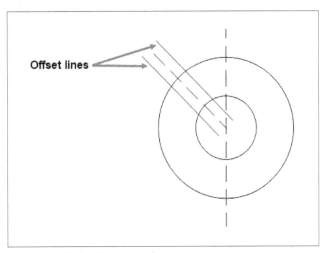

Figure-35. Offset lines created

- Click on the **Trim** tool from **Modify** toolbar and trim the extra entities; refer to Figure-36.

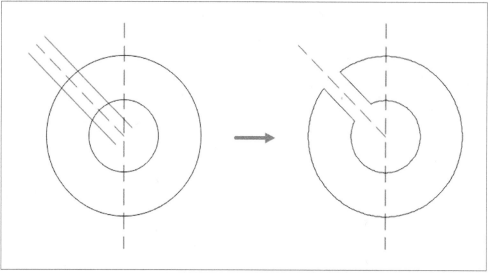

Figure-36. Entities trimmed

- Click on the **Vertical** tool from **Lines** drop-down and create a vertical line of length **2.9652** at the center point of circles; refer to Figure-37.

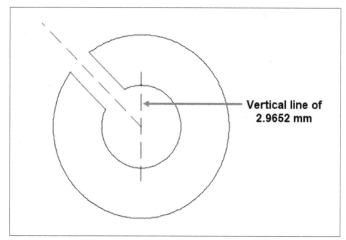

Vertical line of
2.9652 mm

Figure-37. Vertical line created

- Press **Esc** button or click **RMB** to exit the tool.
- Select the first dashed line and click on the **Offset** tool. The **Tool Options** will be displayed and an offset line will be attached to cursor.
- Specify the offset distance as **3** in the **Distance** edit box and create the offset line at the left side.
- Similarly, create the offset lines at a distance of **3.5** and **4**; refer to Figure-38.

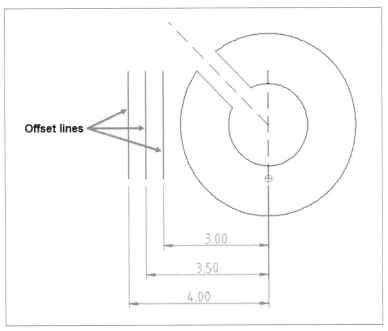

Figure-38. Offset lines created

- Click on the **2 Points** tool from **Line** toolbar and create a line as shown in Figure-39.

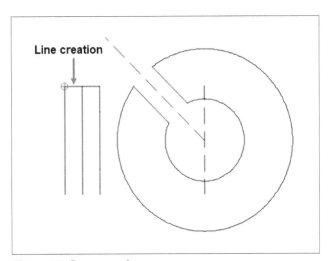

Figure-39. Line created

- Press **Esc** button or click **RMB** to exit the tool.
- Select the recently created line and click on the **Offset** tool from **Modify** toolbar and create offset lines at a distance of **0.6976** and **1.0826**; refer to Figure-40.

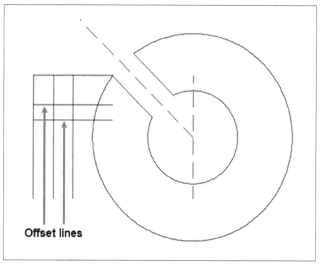

Figure-40. Offset lines created

- Click on the **Mirror** tool from **Modify** toolbar and create mirror entities of recently created lines; refer to Figure-41.

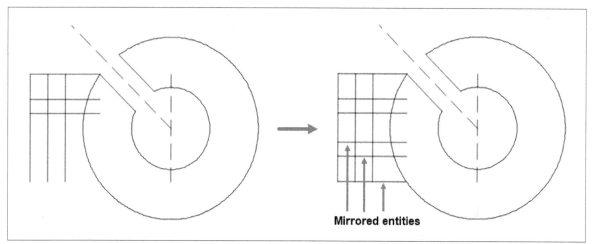

Figure-41. Mirrored entities created

- Click on the **Trim** tool from **Modify** toolbar and trim the extra enities as shown in Figure-42. Note that you can also join the entity using **2 Points** tool if needed.

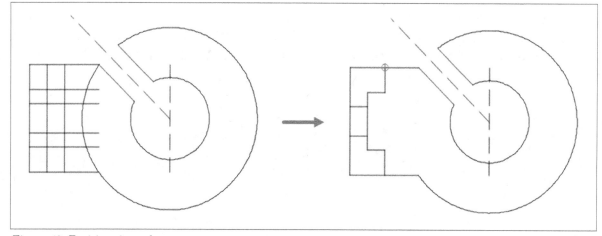

Figure-42. Entities trimmed

- Click on the **Vertical** tool from **Line** toolbar and create a dashed line of length **5.50** at the center point of the circles; refer to Figure-43.

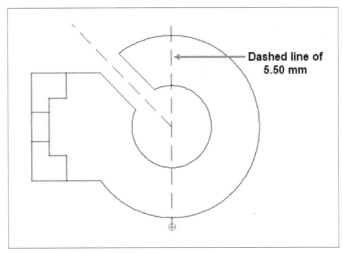

Figure-43. Dashed line created

- Select the recently created dashed line and click on the **Offset** tool from **Modify** toolbar and create offset lines at a distance of **3.80**, **4.50**, **5.00**, and **5.53**; refer to Figure-44.

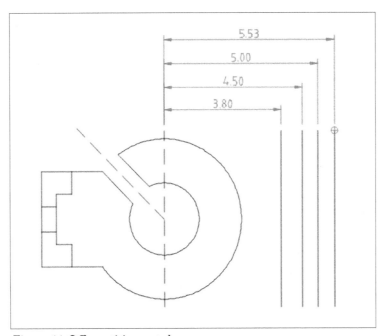

Figure-44. Offset entities created

- Click on the **2 Points** tool from **Line** toolbar and create a line as shown in Figure-45.

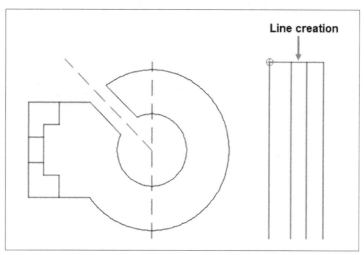

Figure-45. Line created

- Press **Esc** button or click **RMB** to exit the tool.
- Select the recently created line and click on the **Offset** tool from **Modify** toolbar and create offset lines at a distance of **0.25** and **0.50**; refer to Figure-46.

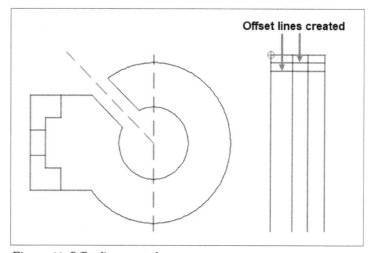

Figure-46. Offset lines created

- Click on the **Mirror** tool from **Modify** toolbar and create mirror entities of recently created lines; refer to Figure-47.

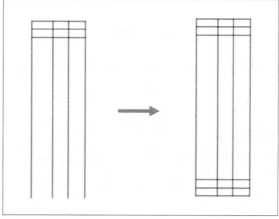

Figure-47. Entities mirrored

- Click on the **Trim** tool from **Modify** toolbar and trim the entities as shown in Figure-48.

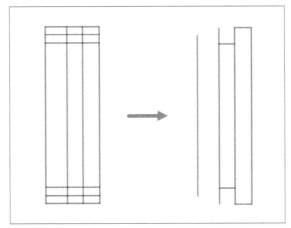

Figure–48. Entities trimmed

- Click on the **3 Points** tool from **Curve** toolbar and create an arc as shown in Figure-49.

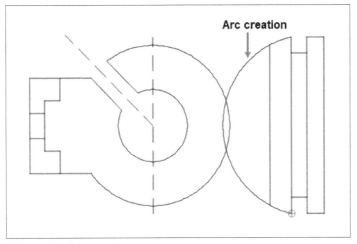

Figure–49. Arc created

- Press **Esc** button or click **RMB** to exit the tool.
- Click on the **Fillet** tool from **Modify** toolbar and apply a fillet of radius **2.5** between the circle and an arc; refer to Figure-50.

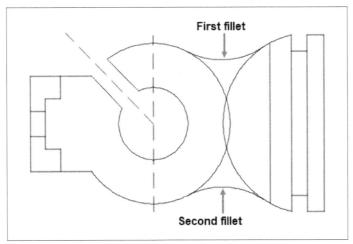

Figure-50. Fillets applied

- Press **Esc** button or click **RMB** to exit the tool.
- Click on the **Trim** tool from **Modify** toolbar and trim the extra entities. The sketch will be created; refer to Figure-51.

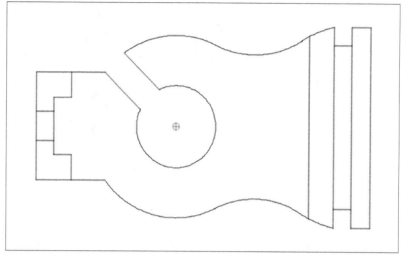

Figure-51. Sketch created

Applying Dimensions

- Click on the **Horizontal** tool from **Dimension** toolbar and apply the horizontal dimensions as shown in Figure-52.

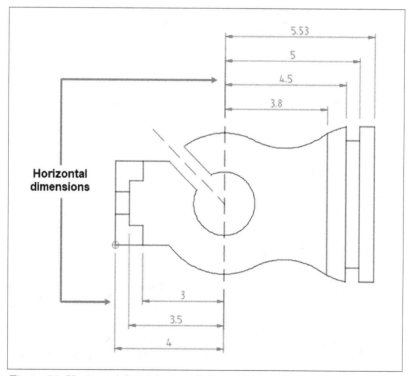

Figure-52. Horizontal dimensions applied

- Press **Esc** button or click **RMB** to exit the tool.
- Click on the **Vertical** tool from **Dimension** toolbar and apply the vertical dimensions as shown in Figure-53.

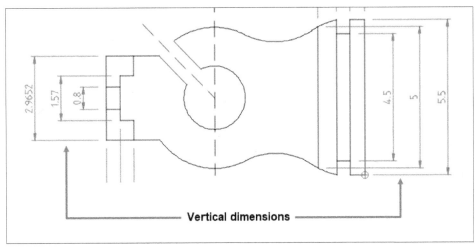

Figure-53. Vertical dimensions applied

- Press **Esc** button or click **RMB** to exit the tool.
- Click on the **Aligned** tool from **Dimension** toolbar and apply the aligned dimension as shown in Figure-54.

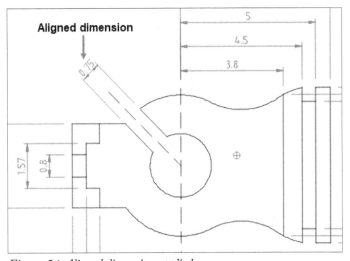

Figure-54. Aligned dimension applied

- Press **Esc** button or click **RMB** to exit the tool.
- Click on the **Angular** tool from **Dimension** toolbar and apply the angular dimension as shown in Figure-55.

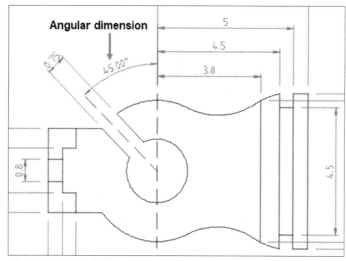

Figure-55. Angular dimension applied

- Press **Esc** button or click **RMB** to exit the tool.
- Click on the **Radial** tool from **Dimension** toolbar and apply the radial dimension. The sketch will be created along with the dimensions; refer to Figure-56.

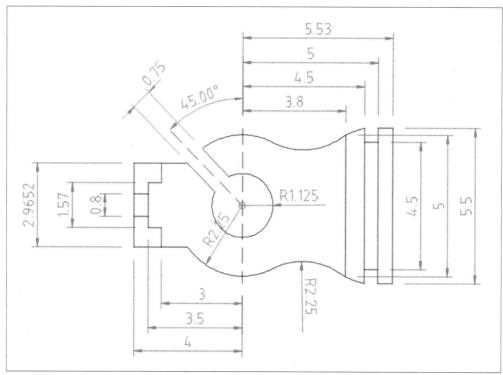

Figure-56. Sketch created along with the dimensions

- Press **Esc** button or click **RMB** to exit the tool.

PRACTICAL 4

In this practice session, you will create sketches of different views for the drawing given in Figure-57. Note that all the dimensions are in **inch**.

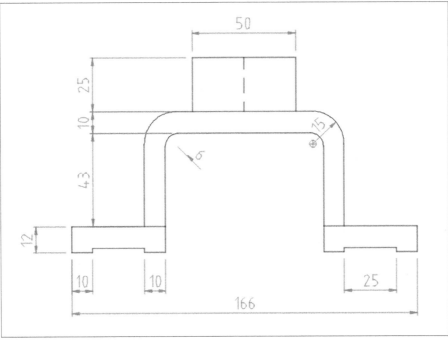

Figure-57. Practical 4

Starting a New Sketch

- Start LibreCAD if not started yet.
- Click on the **Vertical** tool from **Line** toolbar and create a vertical line of length **25**; refer to Figure-58.

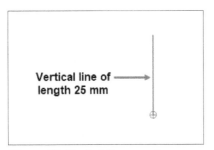

Figure-58. Vertical line created

- Press **Esc** button or click **RMB** to exit the tool.
- Select recently created line and click on the **Offset** tool from **Modify** toolbar.
- Create an offset line at a distance of **50**; refer to Figure-59.

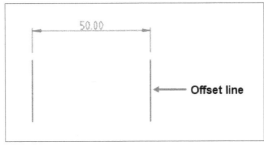

Figure-59. Offset line created

- Click on the **2 Points** tool from **Line** toolbar and join the recently created lines; refer to Figure-60.

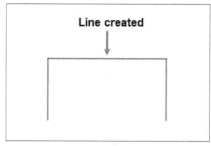

Figure-60. Line created

- Press **Esc** button or click **RMB** to exit the tool.
- Click on the **Vertical** tool from **Line** toolbar and select **Dash (small)** option from **Linetype** drop-down.
- Select **Snap Middle** option from **Snap Selection** toolbar and create a line of length **25** at the mid point of recently created line; refer to Figure-61.

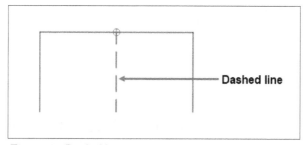

Figure-61. Dashed line created

- Press **Esc** button or click **RMB** to exit the tool.
- Click on the **Horizontal** tool from **Line** toolbar and select **By Layer** option from **Linetype** drop-down.
- Create a line of length **96** at the mid point of dashed line; refer to Figure-62.

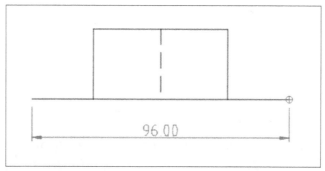

Figure-62. Line created

- Press **Esc** button or click **RMB** to exit the tool.
- Select the recently created line and click on the **Offset** tool from **Modify** toolbar.
- Create an offset line at a distance of **10**; refer to Figure-63.

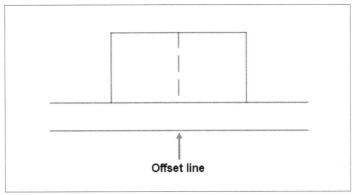

Figure-63. Offset line created

- Click on the **Vertical** tool from **Line** toolbar and select **End** option from **Snap Point** drop-down.
- Create a line of length **53** at both the ends of **96** line; refer to Figure-64.

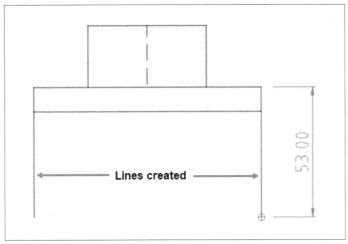

Figure-64. Lines created

- Press **Esc** button or click **RMB** to exit the tool.
- Select the recently created line and click on the **Offset** tool from **Modify** toolbar.
- Create the lines at an offset distance of **10** from both the lines of length **53**; refer to Figure-65.

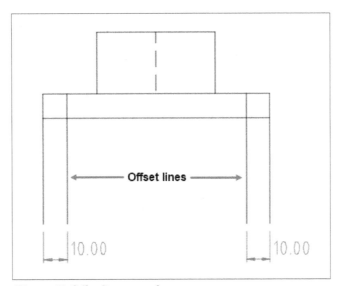

Figure-65. Offset lines created

• Click on the **Trim Two** tool from **Modify** toolbar and trim the extra entities as shown in Figure-66.

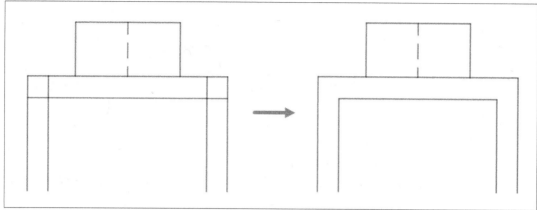

Figure-66. Entities trimmed

• Press **Esc** button or click **RMB** to exit the tool.
• Click on the **Fillet** tool from **Modify** toolbar and select the **Trim** check box and specify the radius of fillet as **6** in the **Radius** edit box of **Tool Options**.
• Apply the fillet feature on the entities as shown in Figure-67.
• Similarly, apply the fillet of radius **15** on the entities as shown in Figure-67.

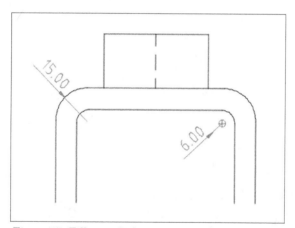

Figure-67. Fillets applied

• Press **Esc** button or click **RMB** to exit the tool.
• Create the sketch using **Horizontal** and **Vertical** tools from **Line** toolbar as shown in Figure-68.

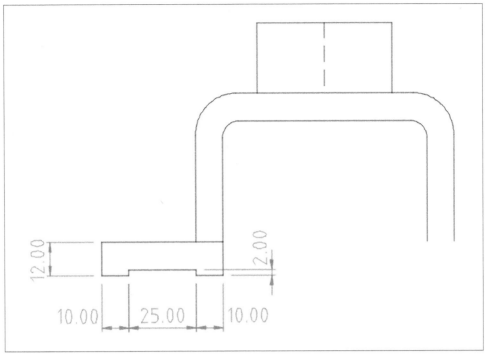

Figure-68. Sketch created

- Click on the **Mirror** tool from **Modify** toolbar and create the mirrored entities of recently created sketch; refer to Figure-69.

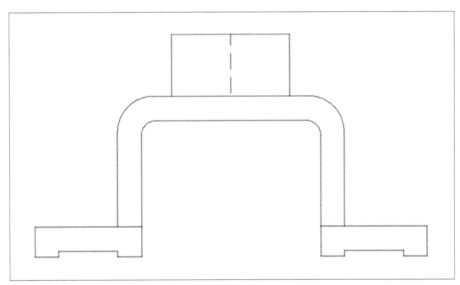

Figure-69. Mirrored entities created

Applying Dimensions

- Click on the **Horizontal** tool from **Dimension** toolbar and apply the horizontal dimensions as shown in Figure-70.

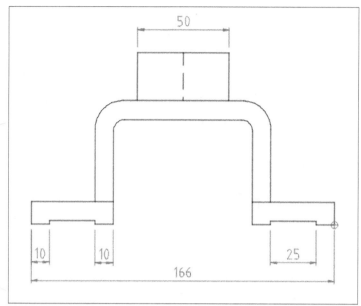

Figure-70. Horizontal dimensions applied

- Press **Esc** button or click **RMB** to exit the tool.
- Click on the **Vertical** tool from **Dimension** toolbar and apply the vertical dimensions to the sketch as shown in Figure-71.

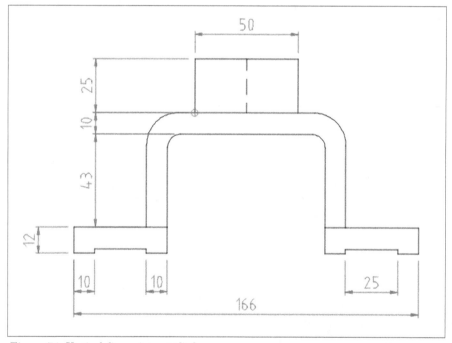

Figure-71. Vertical dimensions applied

- Press **Esc** button or click **RMB** to exit the tool.
- Click on the **Radial** tool from **Dimension** toolbar and apply the radial dimensions to the sketch. The sketch will be created along with the dimensions; refer to Figure-72.

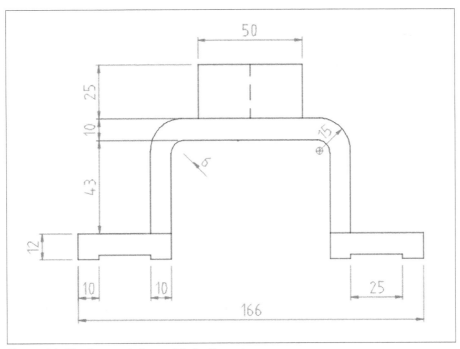

Figure-72. Sketch created along with the dimensions

- Press **Esc** button or click **RMB** to exit the tool.

PRACTICAL 5

Create the sketch as shown in Figure-73. Also, dimension the sketch as per the figure. Note that all the dimensions are in **inch**.

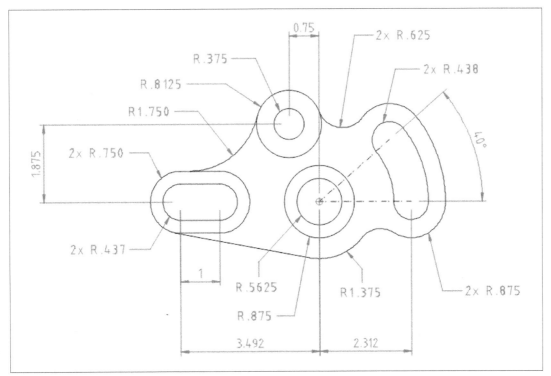

Figure-73. Practical 5

Starting a New Sketch

- Start LibreCAD if not started yet.

- Click on the **Center, Radius** tool from **Circle** toolbar and create a circle of radius **1.375**, **0.875**, and **0.5625** at the same center; refer to Figure-74.

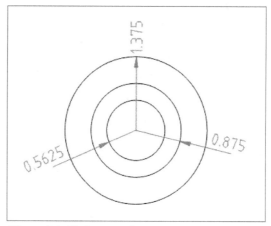

Figure-74. Circles created

- Press **Esc** button or click **RMB** to exit the tool.
- Click on the **Horizontal** tool from **Line** toolbar and create the dashed horizontal lines of length **3.492** and **2.312** from the center of circles; refer to Figure-75.

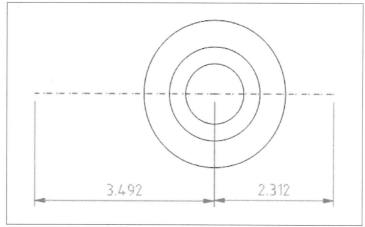

Figure-75. Dashed lines created

- Press **Esc** button or click **RMB** to exit the tool.
- Select the recently created **3.492** dashed line and click on the **Offset** tool from **Modify** toolbar.
- Create a dashed offset line at a distance of **1.875**; refer to Figure-76.

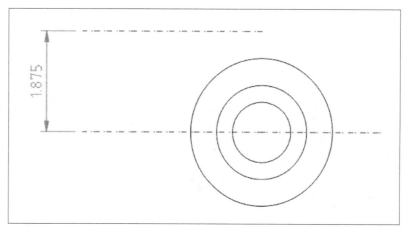

Figure-76. Offset line created

- Click on the **Vertical** tool from **Line** toolbar and create a dashed vertical line joining the horizontal lines; refer to Figure-77.

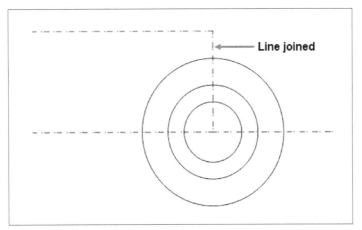

Figure-77. Horizontal lines joined

- Press **Esc** button or click **RMB** to exit the tool.
- Select recently created vertical line and click on the **Offset** tool from **Modify** toolbar.
- Create a dashed offset line at a distance of **0.750**; refer to Figure-78.

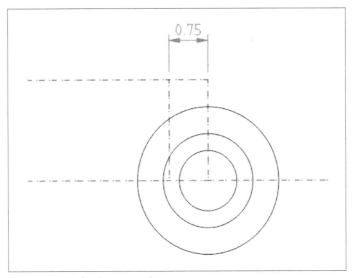

Figure-78. Offset line created

- Click on the **Center, Radius** tool from **Circle** toolbar and create the circles of radius **0.8125** and **0.375** at the same center; refer to Figure-79.

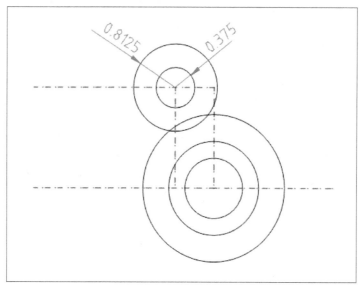

Figure-79. Circles created

- Press **Esc** button or click **RMB** to exit the tool.
- Click on the **Center, Radius** tool from **Circle** toolbar and create the circles of radius **0.750** and **0.437** at the same center; refer to Figure-80.

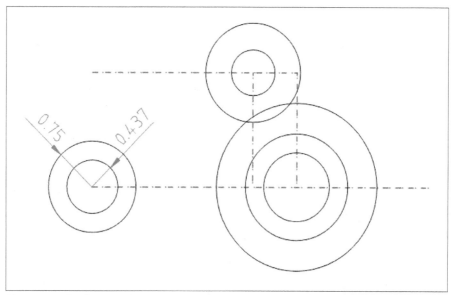

Figure-80. Circles created

- Press **Esc** button or click **RMB** to exit the tool.
- Click on the **Horizontal** tool from **Line** toolbar and create a dashed line at a distance of **1** from the center of recently created circles; refer to Figure-81.

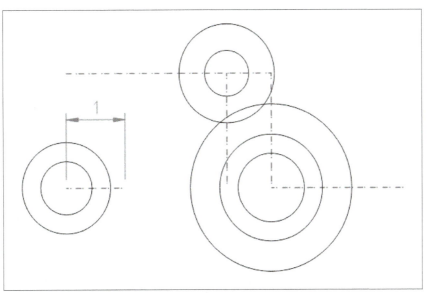

Figure-81. Dashed line created

- Press **Esc** button or click **RMB** to exit the tool.
- Click on the **Move/Copy** tool from **Modify** toolbar and copy the recently created circles to the endpoint of recently created **1** line; refer to Figure-82.

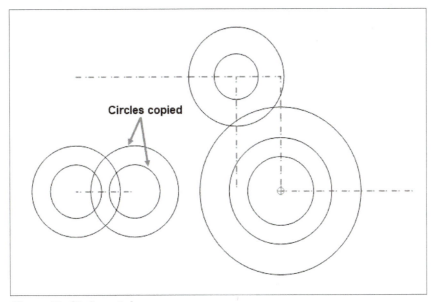

Figure-82. Circles copied

- Click on the **Tangent (C,C)** tool from **Line** toolbar and create tangent lines using the recently created circles; refer to Figure-83.

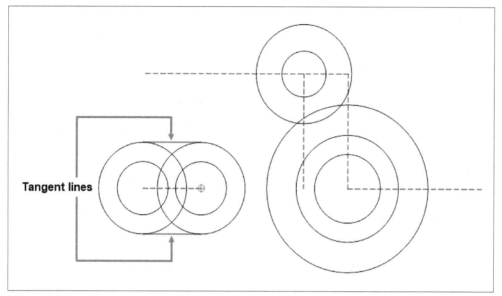

Figure-83. Tangent lines created

- Press **Esc** button or click **RMB** to exit the tool.
- Click on the **Center, Radius** tool from **Circle** toolbar and create the circles of radius **0.875** and **0.438** at the endpoint of **2.312** dashed line; refer to Figure-84.

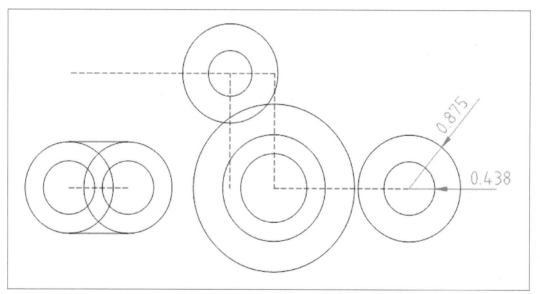

Figure-84. Circles created

- Press **Esc** button or click **RMB** to exit the tool.
- Click on the **Angle** tool from **Line** toolbar and create a dashed line of length **2.312** at an angle of **40°** from the horizontal dashed line of length **2.312**; refer to Figure-85.

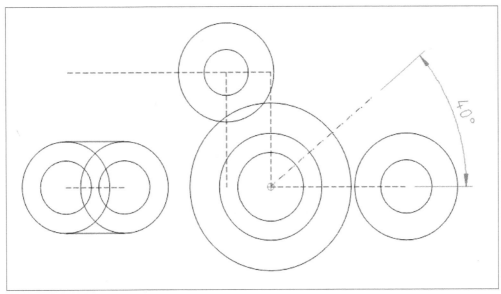

Figure-85. Angled line created

- Press **Esc** button or click **RMB** to exit the tool.
- Click on the **Move/Copy** tool from **Modify** toolbar and copy the recently created circles at the endpoint of angled line; refer to Figure-86.

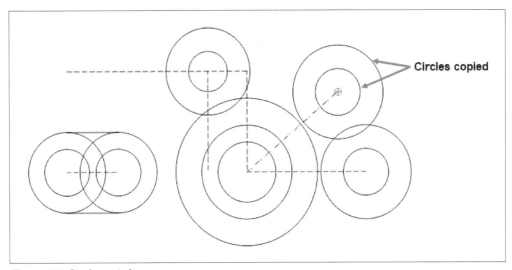

Figure-86. Circles copied

- Click on the **Center, Point, Angles** tool from **Curve** toolbar and create the dashed curve between the recently created circles; refer to Figure-87.

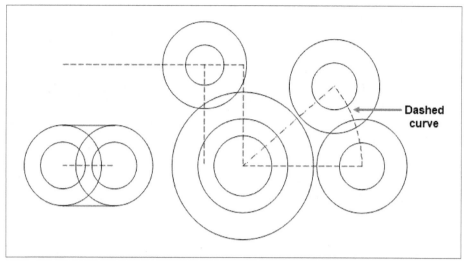

Figure-87. Dashed curve created

- Press **Esc** button or click **RMB** to exit the tool.
- Select the recently created curve and click on the **Offset** tool from **Modify** toolbar and offset the entity at a distance of **0.438** and **0.875**; refer to Figure-88.

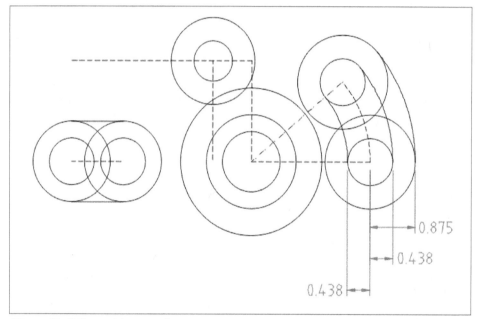

Figure-88. Offset entities created

- Click on the **Fillet** tool from **Modify** toolbar and apply the fillets of radius **1.750** and **0.625** as shown in Figure-89.

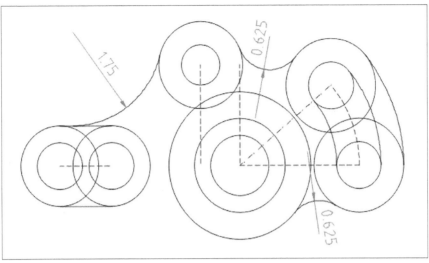

Figure-89. Fillets applied

- Press **Esc** button or click **RMB** to exit the tool.
- Click on the **Tangent (C,C)** tool from **Line** toolbar and create the tangent line between two circles; refer to Figure-90.

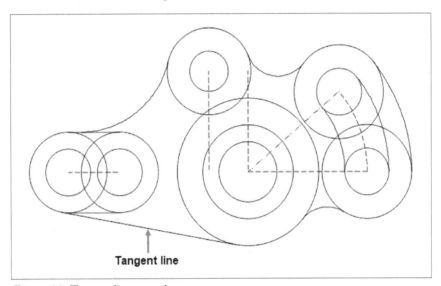

Tangent line

Figure-90. Tangent line created

- Press **Esc** button or click **RMB** to exit the tool.
- Click on the **Tangent (C,C)** tool and create the tangent lines between the circles as shown in Figure-91.

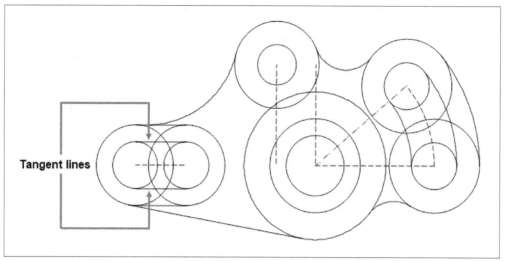

Figure-91. Tangent lines created

- Press **Esc** button or click **RMB** to exit the tool.
- Click on the **Trim** tool from **Modify** toolbar and trim the extra entities. The sketch will be created; refer to Figure-92.

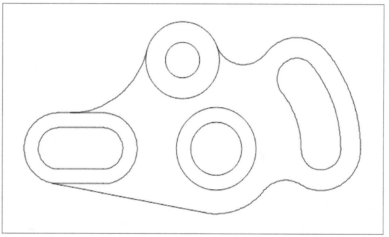

Figure-92. Sketch created

Applying Dimensions

- Click on the **Horizontal** tool from **Dimension** toolbar and apply the horizontal dimensions; refer to Figure-93.

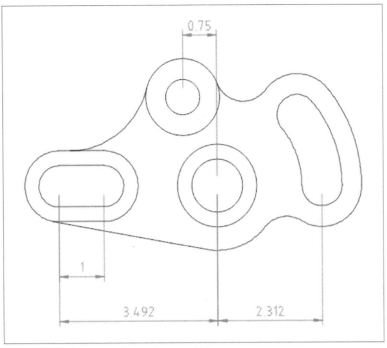

Figure-93. Horizontal dimensions applied

- Press **Esc** button or click **RMB** to exit the tool.
- Click on the **Vertical** tool from **Dimension** toolbar and apply the vertical dimensions; refer to Figure-94.

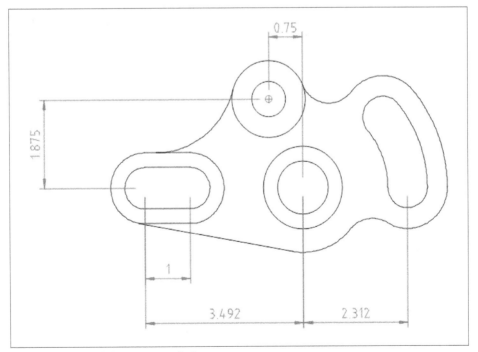

Figure-94. Vertical dimension applied

- Press **Esc** button or click **RMB** to exit the tool.
- Apply the radial dimensions using the **Leader** tool from **Dimension** toolbar and **Text** tool; refer to Figure-95.

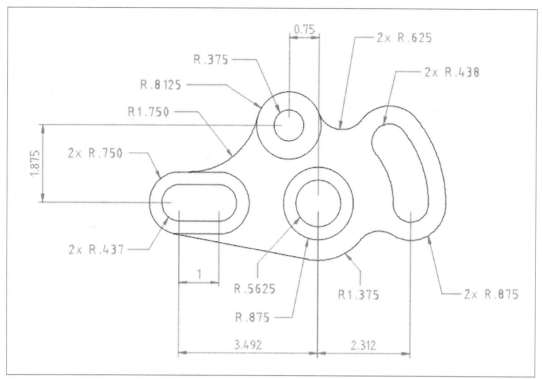

Figure-95. Radial dimensions applied

- Press **Esc** button or click **RMB** to exit the tool.
- Click on the **Angular** tool from **Dimension** toolbar and apply the angular dimension. The sketch will be created along with the dimensions; refer to Figure-96.

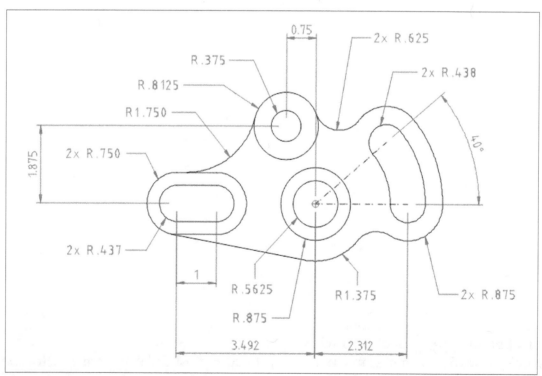

Figure-96. Sketch created along with the dimensions

- Press **Esc** button or click **RMB** to exit the tool.

PRACTICAL 6

Create the sketch as shown in Figure-97. Also, dimension the sketch as per the figure. Note that all the dimensions are in **inch**.

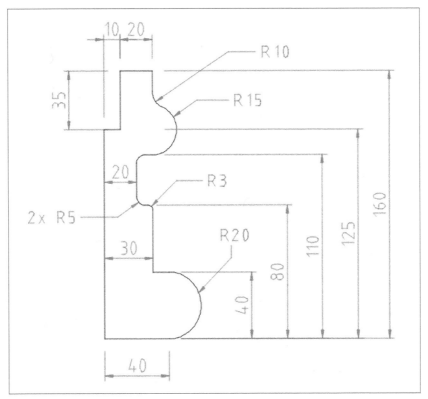

Figure-97. Practical 6

Starting a New Sketch

- Start LibreCAD if not started yet.
- Click on the **Vertical** tool from **Line** toolbar and create the vertical line of length **160**; refer to Figure-98.

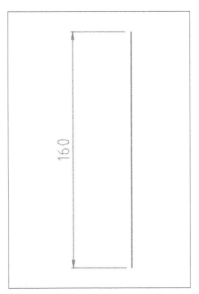

Figure-98. Vertical line created

- Press **Esc** button or click **RMB** to exit the tool.
- Select the recently created vertical line and click on the **Offset** tool from **Modify** toolbar and create the offset entities at a distance of **10**, **20**, **30**, and **40**; refer to Figure-99.

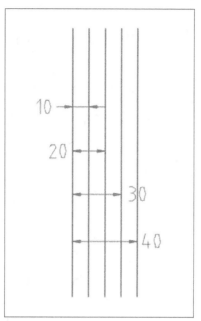

Figure-99. Offset entities created

- Click on the **2 Points** tool from **Line** toolbar and create the horizontal line as shown in Figure-100.

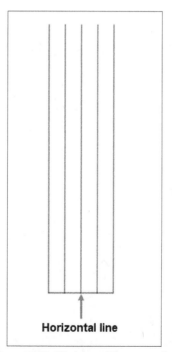

Horizontal line

Figure-100. Horizontal line created

- Press **Esc** button or click **RMB** to exit the tool.
- Select the recently created horizontal line and click on the **Offset** tool from **Modify** toolbar and create the offset entities at a distance of **40**, **80**, **110**, **125**, and **160**; refer to Figure-101.

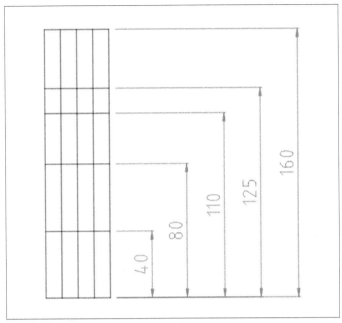

Figure-101. Offset entities created

- Click on the **2 Points, Circle** tool from **Circle** toolbar and create a circle of radius **20**; refer to Figure-102.

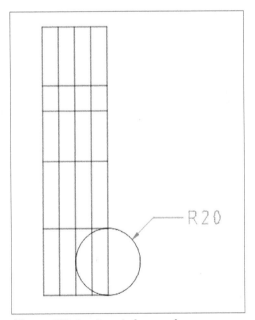

Figure-102. 2 points circle created

- Press **Esc** button or click **RMB** to exit the tool.
- Click on the **Center, Radius** tool from **Circle** toolbar and create a circle of radius **15**; refer to Figure-103.

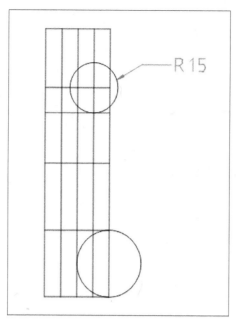

Figure-103. Circle created

- Press **Esc** button or click **RMB** to exit the tool.
- Click on the **Fillet** tool from **Modify** toolbar and apply the fillets of radius **3**, **5**, and **10**; refer to Figure-104.

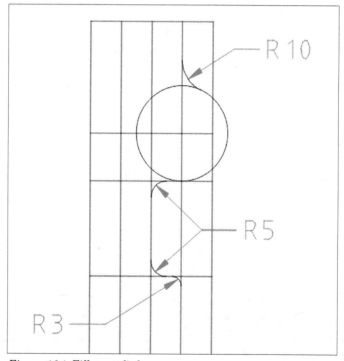

Figure-104. Fillets applied

- Press **Esc** button or click **RMB** to exit the tool.
- Click on the **Trim** tool from **Modify** toolbar and trim the extra entities. The sketch will be created; refer to Figure-105.

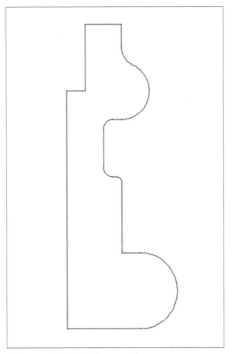

Figure-105. Sketch created

- Press **Esc** button or click **RMB** to exit the tool.

Applying the Dimensions

- Click on the **Vertical** tool from **Dimension** toolbar and apply the vertical dimensions; refer to Figure-106.

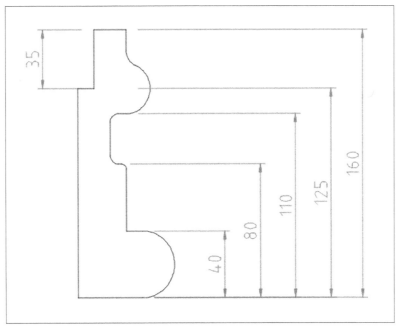

Figure-106. Vertical dimensions applied

- Press **Esc** button or click **RMB** to exit the tool.
- Click on the **Horizontal** tool from **Dimension** toolbar and apply the horizontal dimensions; refer to Figure-107.

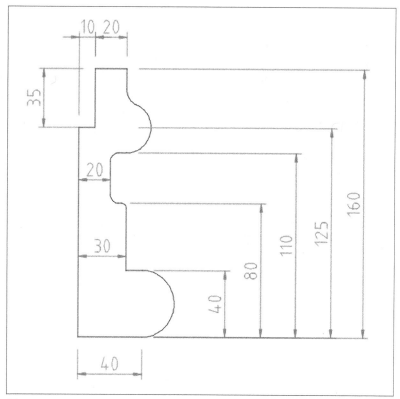

Figure-107. Horizontal dimensions applied

- Press **Esc** button or click **RMB** to exit the tool.
- Apply the radial dimensions using **Leader** tool from **Dimension** toolbar and **Text** tool. The sketch will be created along with the dimensions; refer to Figure-108.

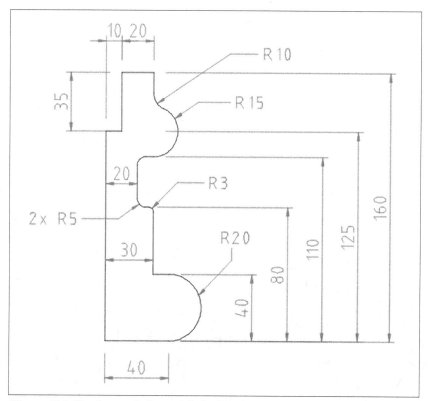

Figure-108. Sketch created along with the dimensions

- Press **Esc** button or click **RMB** to exit the tool.

PRACTICAL 7

Create the sketch as shown in Figure-109. Also, dimension the sketch as per the figure. Note that all the dimensions are in **inch**.

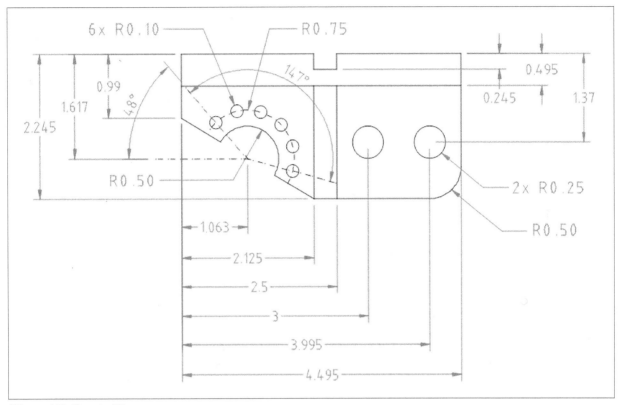

Figure-109. Practical 7

Starting a New Sketch

- Start LibreCAD if not started yet.
- Click on the **Horizontal** tool from **Line** toolbar and create a horizontal line of length **4.495**; refer to Figure-110.

Figure-110. Horizontal line created

- Press **Esc** button or click **RMB** to exit the tool.
- Select the recently created line and click on the **Offset** tool from **Modify** toolbar and create the offset entities at the distance of **0.245**, **0.495**, **0.990**, **1.370**, **1.617**, and **2.245**; refer to Figure-111.

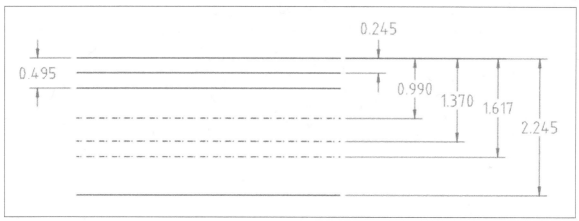

Figure-111. Offset entities created

- Click on the **2 Points** tool from **Line** toolbar and create a line joining the entities; refer to Figure-112.

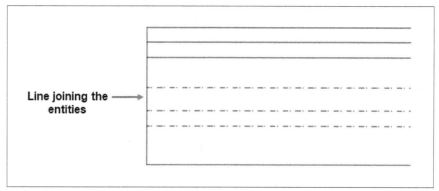

Figure-112. Line created

- Press **Esc** button or click **RMB** to exit the tool.
- Select the recently created 2 point line and click on the **Offset** tool from **Modify** toolbar and create the offset entities at the distance of **1.063**, **2.125**, **2.500**, **3.000**, **3.995**, and **4.495** as shown in Figure-113.

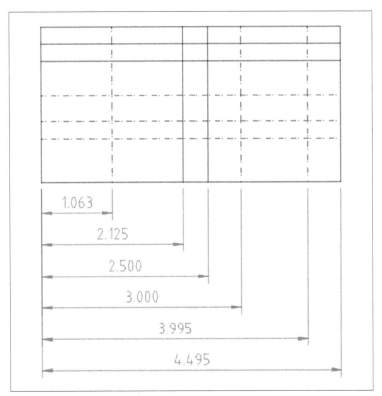

Figure-113. Offset entities created

- Click on the **2 Points** tool from **Line** toolbar and create a line joining the entities; refer to Figure-114.

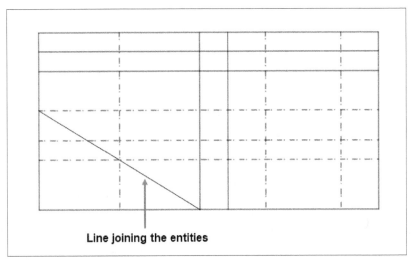

Line joining the entities

Figure-114. Line created

- Press **Esc** button or click **RMB** to exit the tool.
- Click on the **Center, Radius** tool from **Circle** toolbar and create the circles of radius **0.75**, **0.50**, and **0.25**; refer to Figure-115.

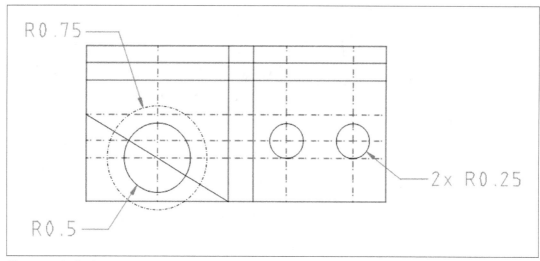

Figure-115. Circles created

- Press **Esc** button or click **RMB** to exit the tool.
- Click on the **Angle** tool from **Line** toolbar and create a line of length **1** and angle **-15°** from the center point of **0.5** radius circle; refer to Figure-116.

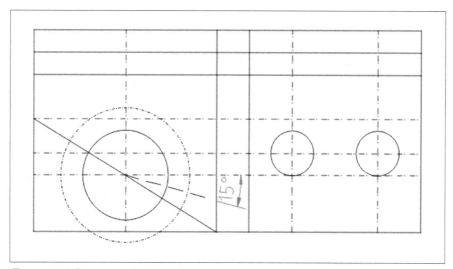

Figure-116. Line created with angle

- Press **Esc** button or click **RMB** to exit the tool.
- Click on the **Center, Radius** tool from **Circle** toolbar and create the circle of radius **0.10** at the intersection point; refer to Figure-117.

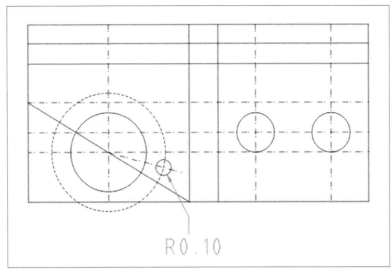

Figure-117. Circle created

- Click on the **Rotate** tool from **Modify** toolbar and create the **5** copies of recently created circle at an angle of **29.4°** each; refer to Figure-118.

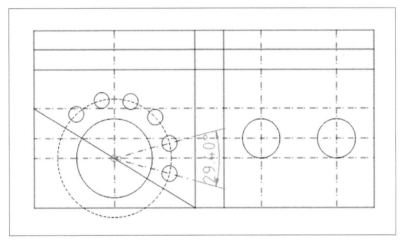

Figure-118. Multiple copies created

- Press **Esc** button or click **RMB** to exit the tool.
- Click on the **Fillet** tool from **Modify** toolbar and apply a fillet of radius **0.50**; refer to Figure-119.

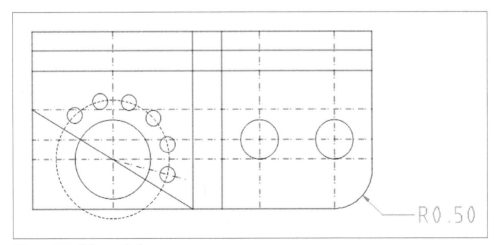

Figure-119. Fillet applied

- Press **Esc** button or click **RMB** to exit the tool.

- Click on the **Trim** tool from **Modify** toolbar and trim the extra entities. The sketch will be created; refer to Figure-120. Note that you can use **2 Points** tool from **Line** toolbar to join the entities.

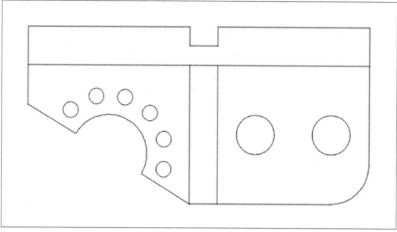

Figure-120. Sketch created

- Press **Esc** button or click RMB to exit the tool.

Applying the dimensions

- Click on the **Horizontal** tool from **Dimension** toolbar and apply the horizontal dimensions; refer to Figure-121.

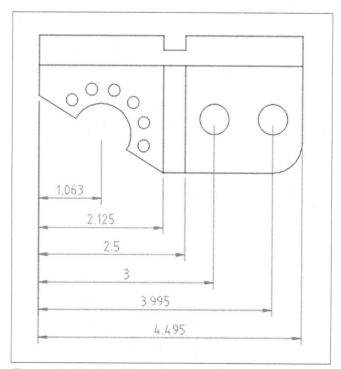

Figure-121. Horizontal dimensions applied

- Press **Esc** button or click **RMB** to exit the tool.
- Click on the **Vertical** tool from **Dimension** toolbar and apply the vertical dimensions; refer to Figure-122.

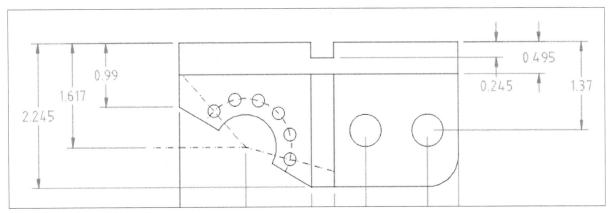

Figure-122. Vertical dimensions applied

- Press **Esc** button or click **RMB** to exit the tool.
- Apply the radial dimensions using **Leader** tool from **Dimension** toolbar and **Text** tool; refer to Figure-123.

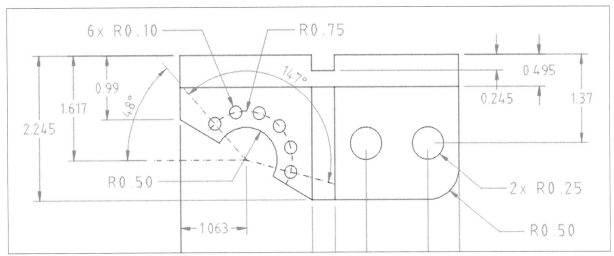

Figure-123. Radial dimensions applied

- Press **Esc** button or click **RMB** to exit the tool.
- Click on the **Angular** tool from **Dimension** toolbar and apply the angular dimensions. The sketch will be created along with the dimensions; refer to Figure-124.

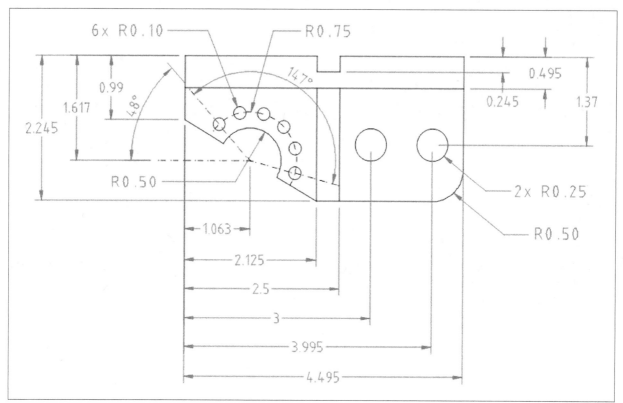

Figure–124. Sketch created along with the dimensions

PRACTICAL 8

Create the sketch as shown in Figure-125. Also, dimension the sketch as per the figure. Note that all the dimensions are in **inch**.

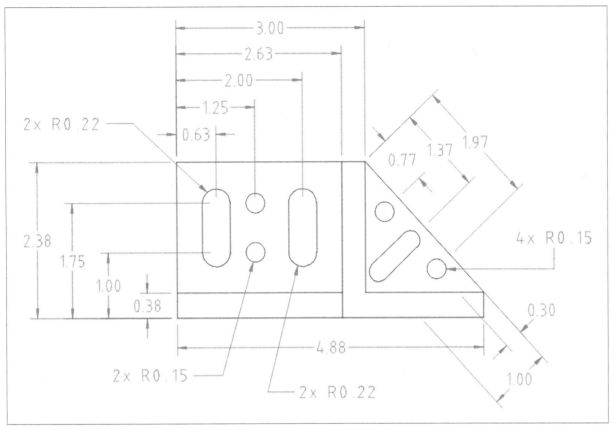

Figure-125. Practical 8

Starting a New Sketch

- Start LibreCAD if not started yet.
- Click on the **Vertical** tool from **Line** toolbar and create a line of length **2.38**; refer to Figure-126.

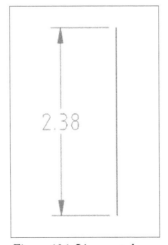

Figure-126. Line created

- Press **Esc** button or click **RMB** to exit the tool.
- Select the recently created line and click on the **Offset** tool from **Modify** toolbar and create an offset entities at a distance of **0.63**, **1.25**, **2.0**, **2.63**, **3.0**, and **4.88**; refer to Figure-127.

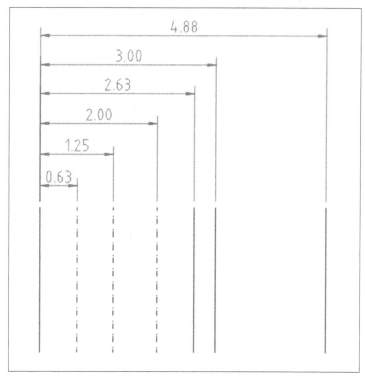

Figure-127. Offset entities created

- Click on the **2 Points** tool from **Line** toolbar and and create a line joining the entities; refer to Figure-128.

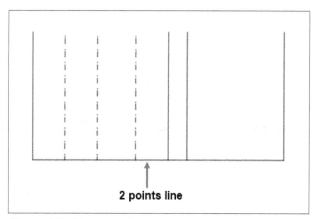

Figure-128. Line created

- Press **Esc** button or click **RMB** to exit the tool.
- Select the recently created 2 points line and click on the **Offset** tool from **Modify** toolbar and create an offset entities at a distance of **0.38**, **1.0**, **1.75**, and **2.38**; refer to Figure-129.

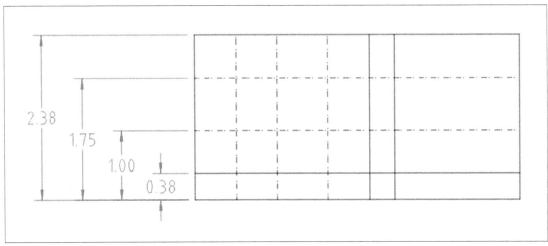

Figure-129. Offset entities created

- Click on the **2 Points** tool from **Line** toolbar and create a line joining the entities; refer to Figure-130.

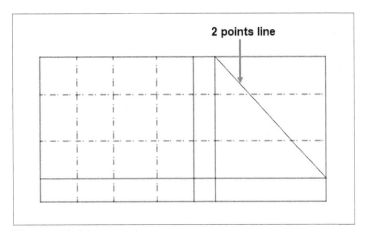

Figure-130. Line created

- Press **Esc** button or click **RMB** to exit the tool.
- Select the recently created line and click on the **Offset** tool from **Modify** toolbar and create the offset entities at a distance of **0.30** and **1.0**; refer to Figure-131.

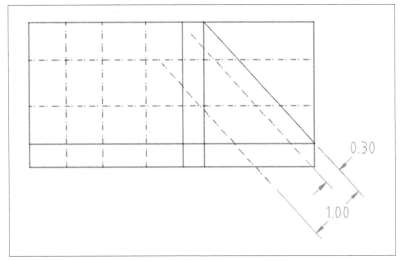

Figure-131. Offset entities created

- Click on the **Relative angle** tool from **Line** toolbar and create a line at an angle of **90°** to the base entity; refer to Figure-132.

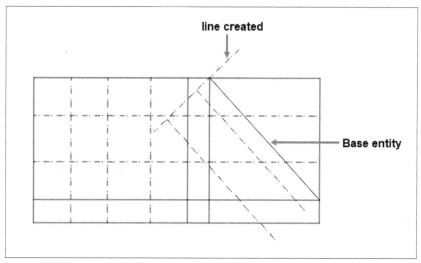

Figure-132. Line created

- Press **Esc** button or click **RMB** to exit the tool.
- Select the recently created line and click on the **Offset** tool from **Modify** toolbar and create an offset entities at a distance of **0.77**, **1.37**, **1.97**; refer to Figure-133.

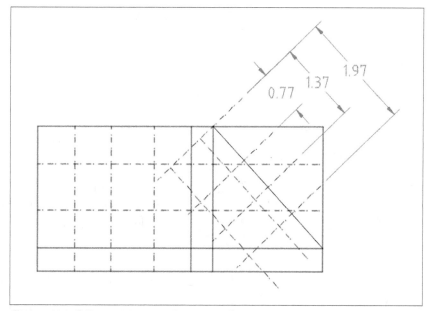

Figure-133. Offset entities created

- Click on the **Center, Radius** tool from **Circle** toolbar and create a circles of radius **0.22** and **0.15**; refer to Figure-134.

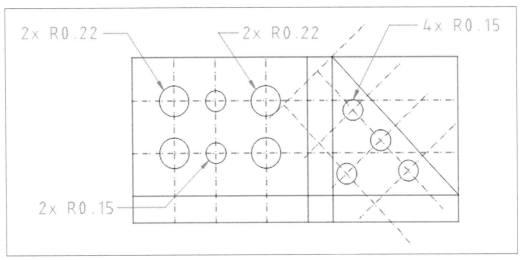

Figure-134. Circles created

- Press **Esc** button or click **RMB** to exit the tool.
- Click on the **2 Points** tool from **Line** toolbar and create lines joining the circles; refer to Figure-135.

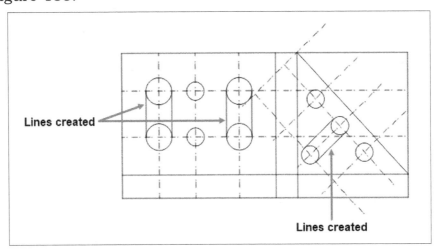

Figure-135. Lines created

- Press **Esc** button or click **RMB** to exit the tool.
- Click on the **Trim** tool from **Modify** toolbar and trim the extra entities. The sketch will be created; refer to Figure-136. Note that you can use **2 Points** tool from **Line** toolbar to join the entities.

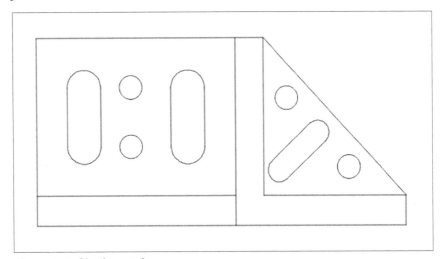

Figure-136. Sketch created

Applying the Dimensions

- Click on the **Vertical** tool from **Dimension** toolbar and apply the vertical dimensions; refer to Figure-137.

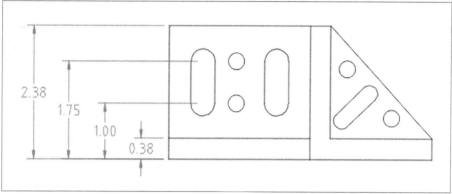

Figure-137. Vertical dimensions applied

- Press **Esc** button or click **RMB** to exit the tool.
- Click on the **Horizontal** tool from **Dimension** toolbar and apply the horizontal dimensions; refer to Figure-138.

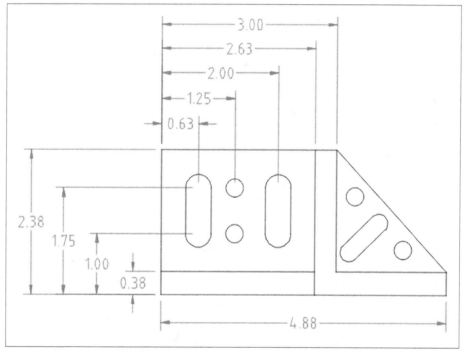

Figure-138. Horizontal dimensions applied

- Press **Esc** button or click **RMB** to exit the tool.
- Click on the **Aligned** tool from **Dimension** toolbar and apply the aligned dimensions; refer to Figure-139.

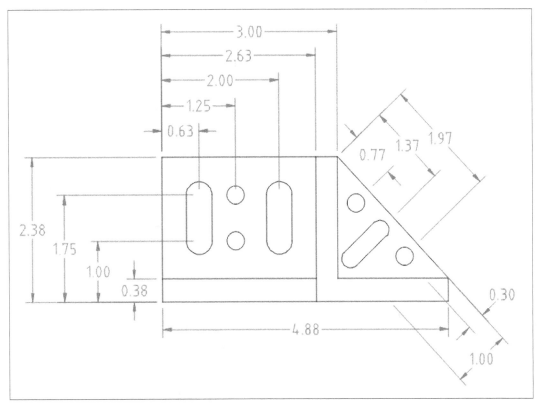

Figure-139. Aligned dimension applied

- Press **Esc** button or click **RMB** to exit the tool.
- Apply the radial dimensions using **Leader** tool from **Dimension** toolbar and **Text** tool. The sketch will be created along with the dimensions; refer to Figure-140.

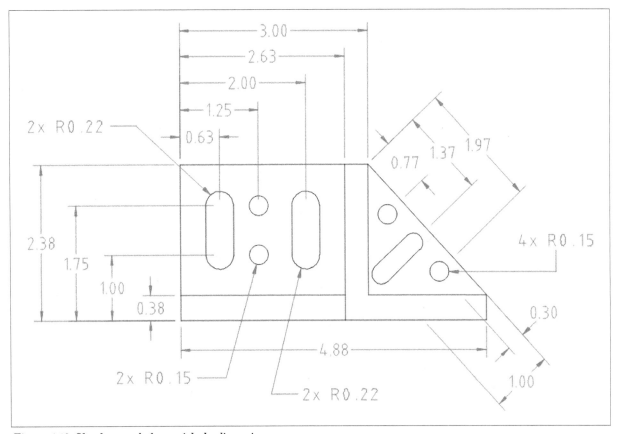

Figure-140. Sketch created along with the dimensions

- Press **Esc** button or click **RMB** to exit the tool.

PRACTICE 1

In this practice session, you will create a sketch for the drawing given in Figure-141. Note that all the dimensions are in **inch**.

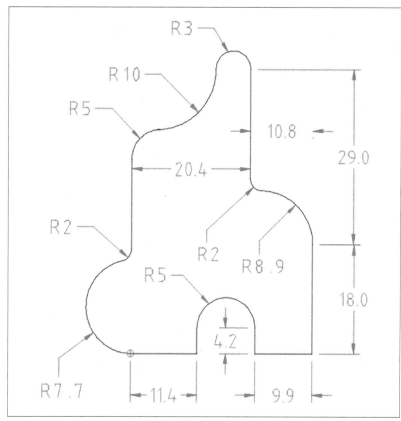

Figure-141. Practice 1

PRACTICE 2

In this practice session, you will create a sketch for the drawing given in Figure-142. Note that all the dimensions are in **inch**.

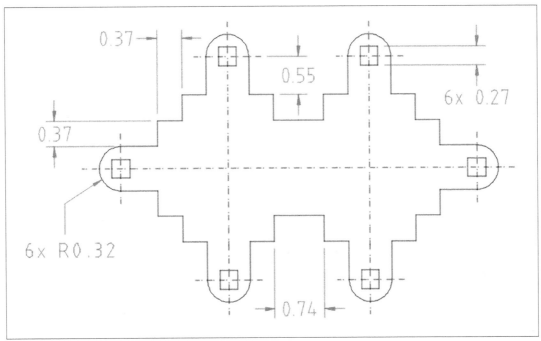

Figure-142. Practice 2

PRACTICE 3

In this practice session, you will create a sketch for the drawing given in Figure-143. Note that all the dimensions are in **inch**.

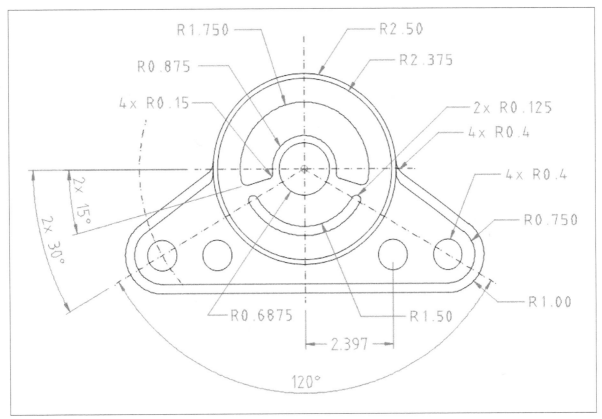

Figure-143. Practice 3

PRACTICE 4

In this practice session, you will create a sketch for the drawing given in Figure-144.
Note that all the dimensions are in **inch**.

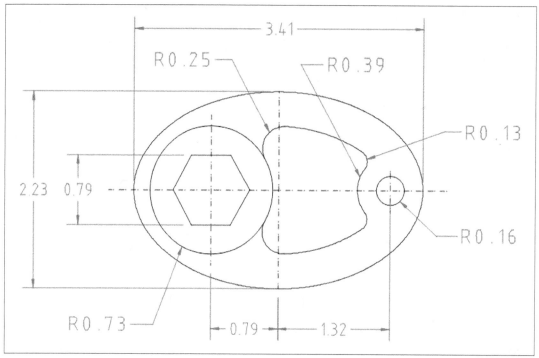

Figure-144. Practice 4

PRACTICE 5

In this practice session, you will create a sketch for the drawing given in Figure-145.
Note that all the dimensions are in **inch**.

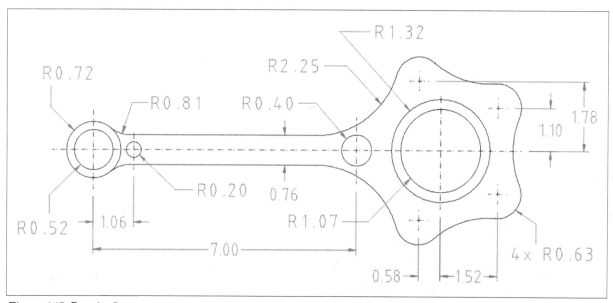

Figure-145. Practice 5

PRACTICE 6

In this practice session, you will create a sketch for the drawing given in Figure-146. Note that all the dimensions are in **inch**.

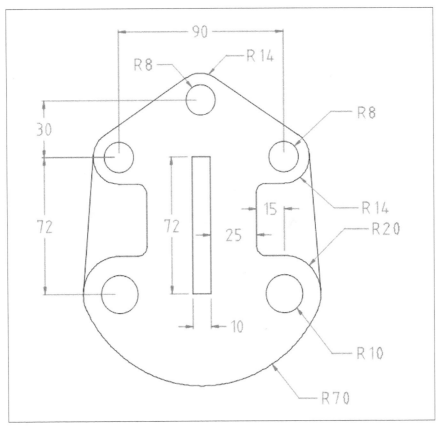

Figure-146. Practice 6

FOR STUDENT NOTES

FOR STUDENT NOTES

Index

OTHER BOOKS BY CADCAMCAE WORKS

Autodesk Revit 2024 Black Book
Autodesk Revit 2023 Black Book
Autodesk Revit 2022 Black Book
Autodesk Revit 2021 Black Book

Autodesk Inventor 2024 Black Book
Autodesk Inventor 2023 Black Book
Autodesk Inventor 2022 Black Book
Autodesk Inventor 2021 Black Book

Autodesk Fusion 360 Black Book (V2.0.15293)

Autodesk Fusion 360 PCB Black Book (V2.0.15509)

AutoCAD Electrical 2024 Black Book
AutoCAD Electrical 2023 Black Book
AutoCAD Electrical 2022 Black Book
AutoCAD Electrical 2021 Black Book

SolidWorks 2023 Black Book
SolidWorks 2022 Black Book
SolidWorks 2021 Black Book

SolidWorks Simulation 2023 Black Book
SolidWorks Simulation 2022 Black Book
SolidWorks Simulation 2021 Black Book

SolidWorks Flow Simulation 2023 Black Book
SolidWorks Flow Simulation 2022 Black Book
SolidWorks Flow Simulation 2021 Black Book

SolidWorks CAM 2023 Black Book
SolidWorks CAM 2022 Black Book
SolidWorks CAM 2021 Black Book

SolidWorks Electrical 2023 Black Book
SolidWorks Electrical 2022 Black Book
SolidWorks Electrical 2021 Black Book

SolidWorks Workbook 2022

Mastercam 2024 for SolidWorks Black Book
Mastercam 2023 for SolidWorks Black Book
Mastercam 2022 for SolidWorks Black Book
Mastercam 2017 for SolidWorks Black Book

Mastercam 2024 Black Book
Mastercam 2023 Black Book
Mastercam 2022 Black Book
Mastercam 2021 Black Book

Creo Parametric 10.0 Black Book
Creo Parametric 9.0 Black Book
Creo Parametric 8.0 Black Book
Creo Parametric 7.0 Black Book

Creo Manufacturing 10.0 Black Book
Creo Manufacturing 9.0 Black Book
Creo Manufacturing 4.0 Black Book

ETABS V20 Black Book
ETABS V19 Black Book
ETABS V18 Black Book

Basics of Autodesk Inventor Nastran 2024
Basics of Autodesk Inventor Nastran 2022
Basics of Autodesk Inventor Nastran 2020

Autodesk CFD 2023 Black Book
Autodesk CFD 2021 Black Book
Autodesk CFD 2018 Black Book

FreeCAD 0.20 Black Book
FreeCAD 0.19 Black Book
FreeCAD 0.18 Black Book

Page Left Blank Intentionally

www.ingramcontent.com/pod-product-compliance
Lightning Source LLC
LaVergne TN
LVHW081657050326
832903LV00026B/1795